SUBSURFACE DRAINAGE OF VALLEY BOTTOM IRRIGATED RICE SCHEMES IN TROPICAL SAVANNAH

CASE STUDIES OF TIEFORA AND MOUSSODOUGOU IN BURKINA FASO

Amadou Keïta

This research was conducted under the auspices of the SENSE Research
School for Socio-Economic and Natural Sciences of the Environment

SUBSURFACE DRAINAGE OF VALLEY BOTTOM IRRIGATED RICE SCHEMES IN TROPICAL SAVANNAH
CASE STUDIES OF TIEFORA AND MOUSSODOUGOU IN BURKINA FASO

Thesis

submitted in fulfilment of the requirements of
the Academic Board of Wageningen University and
the Academic Board of the UNESCO-IHE Institute for Water Education
for the degree of doctor
to be defended in public
on Thursday, 26 March 2015 at 3:30 p.m.
in Delft, The Netherlands

by

Amadou Keïta,
Born in Bamako, Mali

CRC Press/Balkema is an imprint of the Taylor & Francis Group, an informa business

Published by:
CRC Press/Balkema
PO Box 11320, 2301 EH Leiden, The Netherlands
e-mail: Pub.NL@taylorandfrancis.com
www.crcpress.com – www.taylorandfrancis.com

ISBN 978-1-138-02816-6 (Taylor & Francis Group)
ISBN 978-94-6257-263-8 (Wageningen University)

TABLE OF CONTENTS

DEDICATION

To my wife, Edith, my daughters Sodiya, Hanna and Tinia Willie,
and my son Jackdonni

*"All the laws of physics are wrong, at least in some ultimate details,
though many are awfully good approximations"*
– John Wilder Tukey

ACKNOWLEDGEMENT

This research work was made only possible due to an inestimable support and guidance of my supervising committee. In particular, I would like to mention Prof. Bart Schultz who trusted me since the beginning, in that he accepted and patiently helped me to refine the first proposal of this project. He patiently addressed me very constructive criticism, and suggested ways to improve the manuscripts by taking profit from appropriate reference works related to subsurface drainage. He also suggested funding organisations that can help in the acquisition of valuable appropriate equipments without which science is reduced to its artistic dimension. From the beginning to the end of this thesis, Prof. Schultz patiently checked the progression of my work, in the field and in the text. Words are not enough to express my deep gratitude to him. I am also very grateful to Prof. Hamma Yacouba, who suggested me much appreciated guidance in building up the surveys and the controlled experiments, along with the appropriated measurement devices. Prof. Yacouba encouraged me to gradually publish any useful result of my research project not only in workshops, but also and mostly in formal scientific journals. The training I obtained from this exercise is of inestimable value. Finally, I would like to address a special thank to Dr. Laszlo G. Hayde for his precious support in my perpetual effort to make the results available as much as possible not only to the scientific community, but also to the society. Dr. Hayde helped me a lot when scientific equipment acquisition for Europe to Africa had become challenging.

The multidisciplinary dimension of my research work would have been impossible without the precious assistance of several very available persons. Prof. Dayéri Dianou is one of them. I received invaluable guidance during the whole construction process of the concrete microplots for the designed experiments from Prof. Dianou. Furthermore, his encouragements to take into account the microbiological influence in the geochemical processes taking place during iron toxicity strokes in rice grown soils were decisive. I would like to express my deep gratitude to him. I also address my acknowledgment to the late Dr. Youssouf Dembélé, who suggested me, for the first time, to investigate the issue of iron toxicity in rice fields when I expressed my desire to go beyond engineering activities and enter the challenging, but so useful, area of environmental research. Dr. Dembélé proposed me the site of Tiefora in southern Burkina Faso that I later completed with the site of Moussodougou confronted by much more severe iron toxicity. Moussodougou was proposed by Mr. Siaka Koné of Banfora, who also helped me to carry 25 m^3 of contaminated soil from Moussodougou to Ouagadougou, i.e. over more than 500 km. I would like to say a special thanks to him. Finally, I would like to say thanks to my colleagues Sossou Seyram for his assistance in microbiology and Aminata N'Diaye for her precious checks of my English manuscripts.

The field work and the implementation of the experiments involved in this research were made possible only by the help of several very skilled people. In the irrigation schemes of Tiefora and Moussodougou, while implementing the topographic survey, the soil sampling, hydraulic measurements, discussing with farmers, several persons helped me a lot. Among them, I would like to mention for Tiefora: Issouf Coulibaly (my close assistant in data collection), Sibiri Koné (President of farmer's organisation who facilitated so many activities), Abdoulaye Koné (who helped me a lot in making boreholes), and all the plot holders (who provided me with decisive information). For Moussodougou, I would like to report: Kam Gilbert (my close assistant in data collection), Moussa Traoré (the president of farmers' organization, who facilitated all my operations); Tiémoko Ouattara (who helped me bring about the valley bottom maps and the list of farmers), and all the farmers who so willingly cooperated with me. Parallel to the field work, I performed the designed experiments in Kamboinsé with the unique and permanent assistance of Mr. Emmanuel Zongo. Words will miss me if I wanted to tell Mr. Zongo all my gratitude. Finally the help – either in the field or in the data processing – provided by several of my bachelor and master students is unforgettable. Among them, I would like to keep in memory: Aly Bocoum Diallo, Elvis Yao, Moussa Zou, Héritier Rutabara, Noman Koffi Bienvenue Damien, Aboubacar Soumahoro, Ibrahim Niya, and Justine Rutabagaya.

Finally, my research activities would have been called to die if I had not received the comprehension, the patience and the assistance of my lovely family. I say a special thanks to my wife, Edith Sama-Keïta for – among so many things – having taken care of the children during my acrobatic trips between Burkina Faso and The Netherlands, while making sure the bills were paid to keep the village running. I also tenderly kiss my children for their understanding and cooperative spirit. Sodiya, Hanna, Jackdonni and Tinia Willie, beyond the paternal relationship, you have so often contributed to this work by sending me forgotten files and other technical or administrative documents! Your yielding spirit will not be forgotten.

It has been said that "Money is the sinews of war", it is not less in the field of research. My research activities were principally funded by The Netherlands Organization of International Cooperation in Higher Education (NUFFIC). I cannot tell enough the uncommon insight displayed this institution by allowing the so called mid-carrier professionals – these very ones who are on the verge of key decision-making positions in many developing countries, but who are often denied a PhD fellowship opportunity due to their pretended "non-young" age – to undertake research highly connected to key development challenges in their countries. Without NUFFIC, I would have put this research project in closets. Therefore, may the representatives of NUFFIC receive here the expression of my profound gratefulness. My project also received complementary but invaluable funding from: the International Foundation for Science; Sweden (IFS); the French Development Agency (AFD), the Economic and Monetary Union of West Africa (UEMOA), and the United States Agency for International Development (USAID/HED). I am deeply indebted for their precious support.

SUMMARY

The contrasted global population growth with the multiplication of the constraints to developing new irrigation systems puts a special challenge upon human crop production systems that needs to be taken up. The populations in many countries in Asia, Middle East and Africa are expected to double in the coming 50 years. The experience of the green revolution in Asia – during which 70% of food production increase was provided by irrigated agriculture – shows that there is not only a need to strive to increase such crop production systems, but also to improve the production efficiency of existing ones. In fact, as a much worrying case, rice production in valley bottom irrigated lands of African Tropical Savannah is far to yield the expected amount of cereals. One of the major constraints to this production is iron toxicity subsequent to poor drainage conditions. According to Africa Rice, at least 60% of the Tropical Savannah swampy valley bottoms are affected by different degree of iron toxicity. The yield in many areas drops to zero, leaving behind millions of disappointed and impoverished farmers. Therefore, it is not surprising that there is strong research dynamic – ranging from agronomy to microbiology – that strives to propose alleviating solutions to rice iron toxicity. Because prevalent anoxic conditions in the soil combined with iron reducing bacteria development were found a basic contributing factor to iron toxicity, this research has chosen to investigate subsurface drainage potential contributions to solving this issue.

Two complementary series of operations – designed within two project-components and focused on five basic questions closely related to the contributing factors to iron toxicity development – were performed. In fact, the research project was implemented in two major components: field surveys and designed experiments. The field surveys investigated iron toxicity triggering or aggravating factors such as clay proportions, ferrous ion Fe^{2+} concentration, dissolved oxygen, soil acidity or water management. Drawing profit from the knowledge gained in survey research and literature review, two parallel experiments were designed using concrete microplots on one hand and buckets on the other hand, to statistically ascertain the impact of subsurface on soil acidity and ferrous iron concentration changes. All the operations performed within the two components of this research project endeavoured to answer the following five research questions:

1) how is ferrous iron formed and distributed in soils invaded by iron toxicity?
2) how is clay spread within the valley?
3) how is soil permeability affected by clay distribution in the valley?
4) how can water management help improve soil conditions?
5) what is the impact of subsurface drainage on iron toxicity?

The answers to these research questions – already published or in press – are exposed below, followed by the contribution of this research project in two areas: i) science en engineering, and ii) socio-economy.

Clay and ferrous iron may deposit in strata

High ferrous ion Fe^{2+} concentration, inserted into dense clay strata, constitutes an important threat to rice production in several tropical Savannah valley irrigation schemes of West Africa. Many actions are currently undertaken to alleviate iron toxicity. In this study, we have investigated the presence of clay and ferrous iron stratifications within a typical flood prone valley bottom called Tiefora in Burkina Faso. Taking into account the multiple slopes of the valley, two randomized soil samplings were implemented at various depths. Samples were collected as deep as 500 cm, but especially at 30, 50 and 100 cm. The clay percentage was determined by grain size analysis. Ferrous iron concentrations were obtained through the reflectometric method. The stratifications of clay and ferrous ion Fe^{2+} were checked using statistical hypothesis testing (ANOVA and Welch t-Test). Clay percentage within the first 100 cm top soil – 28.9% – was found twice higher than in the layers underneath. Furthermore, ferrous iron was mainly located in the top 30 cm, with a mean concentration of 994 mg/l. This ferrous iron concentration is much higher than found at depths 50 and 100 cm underneath (73 mg/l), while the pH of all the three layers is almost neutral. This striking stratification suggests several means of alleviating iron toxicity. Among these means, we propose maintaining wet conditions during the growing period in the irrigated lands in combination with leaching by subsurface drainage in the fallow periods.

Iron toxicity risk is higher in single season irrigation schemes

With the aim of finding the geochemical differences and helping to build alleviating strategies against iron toxicity, two hematite dominant valley bottoms irrigated rice soils were investigated in the Tropical Savannah region of Burkina Faso. The first site was Tiefora, a 16 ha modern double-season irrigated rice scheme and moderately affected by iron toxicity (10% of the area with a toxicity score of 4). The second site was Moussodougou, a 35 ha traditional single-season irrigated rice valley-bottom, with 50% facing more severe iron toxicity (score 7). Nine soil extracts were taken from three depths – 30, 50 and 100 cm – i.e. 27 at Tiefora and 27 at Moussodougou. Five techniques were used to measure the data: i) the ferrous iron concentration was determined using a reflectometer, ii) a pH-meter yielded the pH, iii) clay-proportions were obtained by United States Department of Army (USDA) grain size analysis and densitometry, iv) the organic matter was determined by oven drying and v) the dry bulk density was determined by using undisturbed soil samples. Statistical hypothesis testing of One-way ANOVA and Welch t-test were applied to the data to isolate the similarities and the differences between the two sites. A geochemical analysis followed to find the causes of these differences. The results showed that while oxidation of pyrite leads to a simultaneous increase in Fe^{2+} concentrations and acidity in the soils of coastal floodplains and mangroves, the oxidation of hematite in Tropical savannah valley bottoms decreases Fe^{2+} but also increases acidity during the dry season. As a consequence, it was found that the single-season irrigation scheme of Moussodougou is significantly (p-value 0.4%) more acidic (pH 5.7) than the double-season system of Tiefora (pH 6.4) with also 750-1800 mg/l higher ferrous ion Fe^{2+}. The ferrous iron reached 3000 mg/l in some layers in Moussodougou. This result is a justification to modernize traditional single-season spate irrigation schemes into double-season irrigated rice schemes.

Subsurface drainage type depends on clay distribution

Waterlogged valley bottom soils of Tropical Savannah are areas where the richest traditional cropping systems are found, but they also face adverse physical and chemical conditions which can drastically drop rice yield. Subsurface drainage has been used for many areas to alleviate waterlogging. However, this drainage is dependent of clay distribution, type and location. The current research analysed these factors using the case of Tiefora. For this purpose nine boreholes, with depths from 2 to 6 m, were realised. Some 50 samples of soils were extracted at various depths, based on soil changes in texture and colour. These samples underwent grain-size-analysis. A comparative non-linear regression was performed on the clay distribution. Quadratic regression was the most appropriate. Clay proportions were high - 20-30% in the 2 m topsoil - in the upstream and middle areas. A more important - 30-40% - peak was reached in the downstream area at 1 m-depth, with a much smaller thickness (less than 50 cm) and higher permeability. These results suggest the application of mole drainage in the valley, except downstream where the classical Hooghoudt pipe subsurface drainage can be implemented.

Subsurface drainage cost can be reduced by taking into account permeability distribution in valley

In flood prone Tropical Savannah valley soils very low infiltration rates often result in acidic conditions favourable to high concentrations of metallic ions, toxic for rice. The infiltration rate determination is important in drainage design to reclaim degraded soils. Several studies have addressed the mapping of the infiltration rate. Yet its relationship with the toposequence of the valley is not clarified. This research has investigated such possibility, examining the case of the irrigated rice valley of Tiefora. Nine boreholes – 1 to 5 m deep – were implemented from upstream to downstream. The Lefranc permeability test of under phreatic conditions in waterlogged soils – used when the impervious layer is close to soil surface or absent – was conducted. First, a comparative regression was applied to the data, including all the parameters of the regression curves. In case of dissimilarity of the infiltration processes, the comparison focused on the final permeability. Our results show a permeability increase from upstream (0.10 ± 0.10 cm/hr) to downstream (greater than 20.0 ± 10.0 cm/h in some places). Taking into account such permeability increase in subsurface drainage system design would result in the implementation of more efficient and cost effective systems.

Data based water management can help to reduce water losses and solve water inequity frictions between farmers

Surface irrigation represents more than 99% of the irrigated area in West Africa and generally includes valley bottoms dedicated to irrigated rice production, which are often denounced as water overusing schemes. Surprisingly, there is neither follow up nor analysis of the irrigation water used in these gravity irrigation systems. Such a work was carried out in the case of the 16 ha Tropical Savannah irrigated rice valley bottom scheme of Tiefora. Using the flow equation of the concrete weir at the headwork, daily water use volumes were calculated as time series covering more than one-year period. The moving average trend analysis reveals that during both the rainy season (1200 mm of rainfall) and the dry season (no rainfall), the main canal gate is almost never closed, keeping a minimum discharge of 200 m^3/day for 4 ha (50 mm/day versus. a local evapotranspiration of 7 mm/day). That stresses the necessity of a more rigorous water management. Furthermore, the autocorrelation analysis by using the ARIMA model showed that the irrigation cycle that ensures equity in water distribution among farm plots is 20 days instead of five. The knowledge of this fact can defuse potential conflicts about equity among farmers: the lack of water in day 4 may be compensated later during the 20-day cycle. It appeared that a simple water level measuring device – installed at the headwork of the main irrigation canal – can produce a time series to which autoregressive moving average model can be applied to yield, at low cost, a thorough assessment of water management in this surface irrigation system.

Subsurface drainage alleviates iron toxicity in mean and long run

Iron toxicity is one of the most important constraints that hinder rice productivity in Tropical Savannah valley bottom irrigated fields, but fortunately that can be alleviated. A too high ferrous iron level in the soil can nullify rice yield. Several research fields – agronomy, pedology through microbiology – strive to provide a solution to this issue. Up to date, the contribution of hydraulics to tackle iron toxicity remained limited. The current research addressed this aspect through controlled experiments on highly ferrous iron contaminated rice hematite soils. Twelve concrete microplots and eight buckets were used to implement two independent designed experiments during a period of 86 days. Drainage and liming were the two factors whose impacts were investigated. Drainage was used with two treatment conditions: 0 mm/day and = 10 mm/day, and liming also had two treatment conditions: *Lime-* = 0 kg/m² and *Lime+* = 1 kg/m² per unit increment increase of the *pH*. Four different responses in the soil were measured: ferrous ion concentration Fe^{2+}, *pH*, oxido reduction potential, and the dissolved oxygen. For the rice, toxicity scores of the International Rice Research Institute were followed up. The results indicate an increase of Fe^{2+} from 935 mg/l to more than 1106 mg/l (at 95% of confidence level), but, which is interesting, with a significant decrease of soil acidity from *pH* 5.6 to 7.3 (95% confidence level). Liming had the same effect in alleviating the acidity. Reduction processes were not hindered by subsurface drainage since the oxydo reduction potential dropped from 84.6 to 9.2 mV, and dissolved oxygen moved from 1 mg/l to less than 0.1 mg/l. Despite of the reduction of the acidity, with such a high ferrous iron level as 1106 mg/l, the iron toxicity score reached 7 in the twelve microplots and the rice died. Still, the reduction of soil acidity provides a new insight on the hematite soils behaviour, opposite to the acidification with subsurface drainage in coastal floodplains and mangrove pyrite. Furthermore, it will lead to less ferrous iron intake by rice roots and in such perspective improve the rice yield. Finally, though liming can achieve the same result, subsurface drainage takes the advantage when this mineral is not available or is expensive.

Project outputs for Tiefora farmers

From the investigations and their supporting activities, two major benefits were brought to the farmers of Tiefora. First, in order to alleviated iron toxicity – which is much less severe in this place than in Moussodougou – and improve rice yield (less than 4 tons/ha), it would be essential to apply according the norms of the Institute of Environment and Agricultural Research (IN.ERA) the complex fertilizer NPK. However, this application should go along with making well built bunds around the farm plots in order to confine the fertilizer and make the mineral more available for the rice roots. This will invigorate the crop and thus strengthen its resistance to iron toxicity. Secondly, the project handed to the farmers' association of Tiefora three key documents: i) an aerial photo the environment of the valley of Tiefora, including the reservoir, the village, the roads and the irrigated valley, ii) a topographical map of the valley bottom, intended to help in potential engineering works on the irrigation system, and iii) a detailed map of the farm plot system, accompanied with the complete list of the farmers and their farm sizes, and the location of iron intoxicated plots for their daily activities.

Project outputs for Moussodougou farmers

Based on the investigation results and due to the severe iron toxicity in Moussodougou, the project provided several advices and handed some key documents to the farmers. Ferrous iron concentration in the soil of Moussodougou can reach 3000 mg/l in many farm plots with acidity as severe as *pH* 4. Since its incorporation into the soil was found to induce the growth of iron reducing bacteria activity, and given the positive conservation impact of organic matter in lightening the soil structure, the project advised the farmers to reduce its use but not to eliminate it completely. In parallel, farmers would have to use the complex NPK as in Tiefora, according to the norms of IN.ERA, but combine it with a careful erection of plot bunds to make the mineral element more available for the rice. Due to the fact that the current single irrigation season during the year in Moussodougou is an aggravating factor of iron toxicity, the project also introduced to the farmers association its ongoing work of developing sprinkler irrigation from groundwater during the dry season. Finally, the project handed to the farmers' association the same set of documents as in Tiefora, but related to the valley bottom of Moussodougou.

Other social impacts

In an ultimate effort to share the insights gained about the iron toxicity alleviation process, this research project produced and uploaded onto the social media *YouTube* several useful videos. The 15 videos uploaded and accessible for everybody, deal with areas as varied as hydrometrics, microbiology, geochemistry and small scale water saving irrigation equipment assembling at village level (without electricity). Many of these videos were very appreciated by the audience. For example, the video of "Innovative irrigation systems in Sub-Saharan Africa (French)" has been viewed/downloaded 500 times/month. Similarly, the video "How to take a sample of disturbed soil or resting in soil immersed at different depths (English)", was viewed/downloaded some 45 times/month. These two videos were classified "creative common" due to their high potential appropriation by third party video productions. Hence, it is expected that the project will have an even higher social impact in the coming months or years.

1. INTRODUCTION

Though being under the threat of several biotic and abiotic stresses, the Tropical Savannah valley bottoms are an important source of soil and water for agriculture. These areas have been used by local populations for water needed activities even before formal irrigation (Masiyandima et al. 2003). The small size of valley bottoms limits their utilization for crop production to less than 10% of the total crop farming area in West Africa (Oosterbaan et al. 1986), farming families produce various crops in the valley bottoms, ranging from rice during the rainy season to maize and vegetables in the dry season. In fact, the richest and more complex traditional farming systems are found in valley bottoms because of their difference in soils compared with uplands and their smaller sizes inciting to more ingeniousness (Steiner 1998).

The traditional exploitation of valley bottoms by farmers strives essentially to produce rice in often very adverse conditions, using unsophisticated infrastructures. The producers usually do not build any flood protection canal or dike around the site. Hence, the sites remain object to seasonal flooding bringing rich colluviums suitable for crop production (Bronkhorst 2006). However, unlike most of the soils in the Tropical Savannah dominated by the red hues of the hematite, the valley bottom soils – particularly when exposed to several cycles of flooding during the year such as in wet and dry seasons in irrigated fields with high groundwater tables – are predominantly made of gleysoils (Schaetzl and Sharon 2005, Ogban and Babalola 2009) weathered from kaolinite and presenting iron redox processes. They are less exposed to the threat of classical sodium salinity build up than to ferrous iron toxicity, which breaks rice yield (Moormann and Breemen 1978).

1.1. Background and objectives

1.1.1. Problem statement

In the Tropical Savannah, the classic salinity build up (excess of sodium) is not the most threatening consequence of poor drainage, but ferrous or iron toxicity[1]. This toxicity was described as an excessive accumulation of iron in the plant, taking place especially when the soil is not well aerated. It drastically reduces the crop yield, particularly rice. The West Africa Rice Development Association WARDA – now called 'Africa Rice' – estimated that at least 60% of the swampy cultivated inland areas of Africa are affected by varying degrees of iron toxicity (Sahravat et al. 1996). The same association reported that in West Africa, iron toxicity causes up to 12% through to 100% of rice yield drops, depending on its severity and the tolerance of the rice variety.

There is an ongoing difficult and very multidisciplinary fight against iron toxicity, though not really successful up to date. Different cultivars of African species – *Oryza glaberrima*) – such as varieties of Nerica developed by WARDA, were tested

---

[1] The iron toxicity scoring system is based on the extent of bronzing symptoms on rice plant leaves, using a scale of 1 to 9. A score of 1 indicates normal growth and 9 means that most plants are dead or dying (Sahravat et al., 2002).

with some degree of success against drought, aluminium, termites, blast, etc. but none has shown a frank resistance against iron toxicity: iron still accumulates in the roots of the most resistant varieties (Majerus et al. 2007). According to WARDA (Somado et al. 2008), in 2008, Nerica varieties for irrigated lowlands were still going under farmers' level tests through some 20 Sub-Saharan African countries to evaluate their tolerance to abiotic and biotic stresses. As suggested by Audebert et al. (2000) the application of nutrients such as nitrogen N, phosphorus P and zinc Zn may also help in reducing toxicity effect on rice. Various researches have considered the inhibition of the sulphate-reducing bacteria (SRB) activities using chemicals such as chromate and metronidazole, but as Dianou (1993) wrote, 'the unknown effects of these substances either on whole telluric microflora or on the plant limits their use in ecosystems'. Further, some researches have been conducted on subsurface drainage with varying degrees of success in improving rice yield in soils impinged by iron toxicity (Mathew et al. 2001). However following our literature review, it seems that the process of subsurface drainage impact on iron toxicity, and its connections with iron and sulphate bacterial growth has not been searched out (Soares JV and Almeida AC 2001, Dianou 2005). These issues are addressed here by investigating in two typical valley bottom irrigation scheme – Tiefora and Moussodougou – located both in Sudanian climate in Burkina Faso (900-1200 mm of annual rainfall, Figure 1.1). Parallel and in order to ascertain the results of these surveys, designed microplot and bucket experiments were also undertaken.

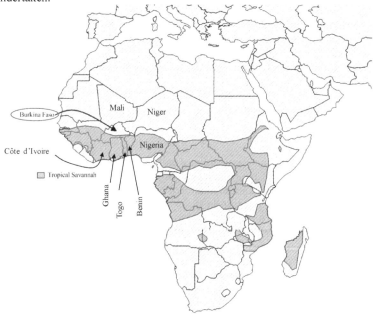

Figure 1.1: Tropical Savannah zone of Africa

The area covers more than 20 countries in Africa, stretching from Senegal to Ethiopia, while expanding toward to south of continent till Madagascar. It covers less than half of Burkina Faso.
Source: adapted from Peel et al. (2007)

In Burkina Faso, the irrigated rice yield remains disappointedly low, sometimes leading farmers to give up their lands. The average yield has been estimated to 2.5-3.0 tons/ha for irrigated rice; this figure being even lower for lowland and rainfed rice fields (Ouattara 1992, Ouédraogo et al. 2005). The production reduction is aggravated by the giving up of some rice fields subsequent to soil degradation due to iron toxicity. For example, Ouattara (1992) reports that 300 ha were abandoned in 1986 at the Valley of Kou (west of the country) due to an excessive level of iron toxicity. In fact this toxicity is a recurrent problem in the Tropical Savannah areas in the south-west and the swampy western regions of Burkina Faso, where the majority of the valleys and lowlands suitable for rice production are located. In short, investigating how subsurface drainage can help improving the rice yield of these areas is the basic problem addressed by this research. The issue is studied in two typical valley bottom irrigation schemes of in Tropical Savannah of Burkina Faso: Tiefora and Moussodougou.

1.1.2. Research questions

It seems that anoxic waterlogging conditions in soils are closely linked to iron toxicity. Many rice production valley bottom fields in Tropical Savannah and Tropical Rainforest (Peel et al. 2007) of West Africa are characterized by a persistent groundwater table. In irrigation schemes, there are some vague ideas about the major contributing factors to swampy conditions. For example, diagnoses conducted in 2009/2010 in two different schemes (Moutori and Tiefora 200 km away in Burkina Faso) – one in a hilly area and located immediately downstream of a dam, and the other located 800 m downstream of the dam – reveal that in both areas, groundwater table is less than 30 cm below surface in the dry season (no rainfall). Irrigation, seepage from irrigation canals and the dam, infiltration from sloping areas, the presence of a relatively shallow impervious layer in the soil might all have an important effect on the phenomenon. On the other hand, research and field observations reveal that there is a direct link between poor drainage and low rice yield and even under strict fertilizer application or with well selected cultivars (Nguu et al. 1988, Wopereis et al. 1999, Becker and Johnson 2001). Farmers' reports show that a decaying pond within a plot due to non-frequent water renewal leads to rice stress and browning. A timely replacement of the pond by freshwater was reported to provide to the rice a clearly visible vitality. However, this regain of vitality would remain limited when the underneath water layers in the soil profile were not replaced. Therefore, the basic question of this research was: *Can subsurface drainage help to improve the rice yield by reducing iron toxicity effect?*

This basic question brings forward the following more specific questions:

1) how is ferrous iron formed and distributed in contaminated soils of irrigated rice valley bottoms?
2) how is clay – an important flow modification factor – spread within the valley?
3) how is soil permeability affected by clay distribution?
4) how can farmers' water management help improve soil conditions?
5) what real impact can subsurface drainage have in alleviating iron toxicity?

1.1.3. Research objectives

Two research components were designed to test the previous five research questions and propose actions to improve rice yield. The first component consisted of field measurements and investigations, while the second was implemented through designed experiments in microplots and buckets. The surveys were implemented in two Tropical Savannah valley bottom irrigation schemes, Tiefora and Moussodougou, both selected for their potential to reveal important factors leading to iron toxicity. Tiefora – a scheme developed at the beginning of years 1960's – was known to present very low rice yield by only slight iron toxicity symptoms. In contrast, Moussodougou was more recent – developed in years 2007-2008 – , having also a very low rice yield of less than 3 tons/ha, but presenting severe iron toxicity symptoms (Sokona et al. 2010, Kanté 2011). In parallel, a second project component made of microplots and bucket experiments was designed according to the principles of Design and Analysis of Experiments – also called DOE – in order to search out subsurface drainage potential impact on ferrous iron concentrations, acidity and other important soil responses variables (Mason et al. 2003, Mathews 2005). Hence, the research focused on testing the five previous hypotheses in these two schemes, with a special interest on what makes the difference between them, and finally what can be done to help farmers achieve much higher rice yields.

1.1.4. Expected results

There is still a strong on-going effort particularly in West Africa to improve the rice growing conditions and yields. Many works are devoted to varietal improvement, the impact of which has indeed been estimated to hundred millions of Euros already in 1998 (Jones et al. 1997, Dalton and Guei 2003). Other works were concentrated of creating bacterial conditions unfavourable to iron toxicity (Ouattara and Jacq 1992, Dianou et al. 1998). While the population growth continues and heightens the needs, irrigated rice yield remains low in Africa due to adverse factors from the ecological environment such as drainage conditions. Therefore, by searching through the means to improve the soil health using drainage (Pretty JN et al. 2003), this work is expected to make a major contribution to a sustainable and effective rice production system in Tropical Savannah. Granted, some researches draw the attention on the unfairness of the Millennium Development Goals (MDGs) initial conditions for Africa and the minimisation of the progress achieved in many areas (Easterly W 2009). Nevertheless, this work intends to contribute to effort to reduce extreme poverty and hunger, not only in Burkina Faso, but also for West African countries covered by Tropical Savannah.

More specifically, the main results expected from this PhD research were:

- Ferrous iron formation and distribution in typical Tropical Savannah irrigated rice valley bottoms hematite soils are identified and described;
- Subsurface drainage impact on the Tropical Savannah valley bottom soils is clearly investigated trough survey and controlled designed experiments
- Methods are found to reduce iron toxicity and improve rice yield in valley bottom irrigated fields of Tiefora and Moussodougou.

1.2.   Scope

The scope of the PhD research is subsurface drainage in relation with iron toxicity, and how this toxicity is impacted by bacterial activity in irrigated valley bottom lands of

Sudanian region. To be valid, measurements obtained from survey and designed experiments had to be processed and analysed with statistical tools (Montgomery 2001). Furthermore, soil medium is complex, resulting for an often long geochemical weathering, and being the home for intense chemical transformations and microbial activities. Therefore, knowledge in the areas of hydraulics, hydrology, geochemistry, chemistry, microbiology, and statistics were to be mobilized to implement this research.

1.3.   Structure of the thesis

The core of this thesis is articulated into eleven main chapters, each one bringing a substantial contribution to the process implemented to provide new insight and propose practical solutions that may contribute to alleviate iron toxicity. In the *first chapter* the special areas that make up valley bottom rice production schemes are described. It naturally draws from this description the key problems that these valleys are facing, and stresses specially iron toxicity issue and the ongoing related research for its alleviation. Then, the chapter proposes subsurface drainage as a mean that possesses the potential to fight against iron toxicity. In the *second chapter*, a literature review is presented. It describes the soil genesis in Tropical Savannah in connection with the climatic and geochemical weathering of minerals. It also stresses what the previous research has brought to light the bacterial interaction with the geochemical processes leading to iron toxicity of rice. Particularly, the role of iron and sulphate reducing bacteria (respectively IRB and SRB) is scrupulously described. Following this literature review, the descriptions of the two valley bottoms investigated – namely Tiefora and Moussodougou – are provided. These descriptions not only cover historical, geographical, climatic, and environmental aspects, but also – by borrowing the knowledge drawn from the valley bottoms characteristics and difficulties as obtained in the first two chapters – lead to a more precise statements of the specific problems met in Tiefora and Moussodougou. These problems are presented of the *third chapter* on "Materials and methods". In order to treat the problems, the research project was divided into two main components, which are also described in this chapter in general terms. In the *forth chapter*, the results of a pre-diagnostic of both Tiefora and Moussodougou sites to update data and problem identification are reported.

   The subsequent five chapters – which all are based on papers published or submitted for publication – describe more thoroughly the groups of research activities undertaken during this project, and their results. In line with the first research question, ferrous distribution and its relationship with iron toxicity in the valley of Tiefora is investigated in *chapter five*. In *chapter six*, this type of investigation is extended to include a comparison between a single irrigation season – namely Moussodougou - and double season irrigation – namely Tiefora – irrigation schemes in terms of ferrous iron and clay (and other soil responses) distribution in a Tropical Savannah valley bottom. Some important theoretical and practical lessons were drawn from this comparative survey study. This chapter also provides an answer to the first research question. In *chapter seven* and *chapter eight*, clay and hydraulic conductivity patterns are investigated within the valley bottom of Tiefora. They provide practical solutions to the drainage issues in the valley bottom of Tiefora, but also give answers respectively to the second and the third research questions. *Chapter nine*, the issue of farmers' water management in connection with equity in distribution and excessive water use leading to water table rising is studied. Time series records of daily water use have been processed using the statistical model ARIMA to bring about the true irrigation cycle and water losses. This chapter provides an answer to the fourth research question about water management. Finally, in *chapter ten*, the investigations with microplots and bucket experiments to statistically assess the impact of subsurface drainage and liming factors on four soil related responses – the ferrous iron, the dissolve oxygen, the *pH*, the oxidoreduction potential – and one rice related response – the iron toxicity score – are reported. This chapter provides a clear answer to the question of the impact of subsurface drainage on iron toxicity and other soil properties. Equally, it solves the fifth research question of this project. It sets clearly how hematite soils behave differently from the pyrite soils found in mangroves and coastal floodplains in terms of ferrous iron and acidity evolution when the soils are subjected to subsurface drainage, bringing the necessity for a change in approach.

The scientific and practical outputs of this research are reported in *chapter eleven*. Both the theoretical and practical findings are reported with a stress on how they contribute to improve our knowledge and methods to fight against iron toxicity in Tropical Savannah irrigated rice valley bottoms. Finally, it is appropriate to remember that this research was implemented with two basic objectives: bringing new insights in iron toxicity issues and propose practical solutions to farmers. For these reasons, it appeared important to meet the farmers, discuss the findings with them and propose short term and mean term solutions. This point is also reported.

2. LITERATURE REVIEW

2.1. Tropical Savannah in Africa

Several classifications of world climate have been proposed and vary in terms of the parameters taken into account and generated results. For example, a system developed in 1931 and modified in 1948 by Thornthwaite (Thornthwaite 1948) proposed a classification that uses four groups of indices to define boundaries between climatic zones: the moisture index, the thermal efficiency index (ratio of total monthly precipitation and total monthly evaporation), the aridity/humidity indices and the index of concentration of thermal efficiency. Although linked to soil conditions through the use of potential evapotranspiration to define atmospheric moisture, this method is currently less used because of its complexity (Crowe 1954, Griffiths 1983) and the unavailability of world maps based upon it. The classification was modified and simplified by Feddema (2005). However, it seems that the most commonly global climate classification used is the one the German botanist and climatologist Köppen (Koppen 1936, Wilcock 1968). The climatic delineation criteria of Köppen's system are based mostly on average monthly and annual values of temperature and precipitation. This system defines 5 major climate types, named from A through to E. Subdivisions in the climate zones are rendered by the adjunction of small letters to the previous capital letters.

The climatic delineation of Köppen was updated by Peel et al. (2007) taking into account more global available data and producing a result that notably shows the importance of Tropical Savannah at global scale, but also in Africa. A total of 1436 precipitations and 331 temperature stations scattered over the African continent were used, with a station concentration for precipitation higher than for temperature, which is more volatile. In this updated delineation, three major climate types are met in Africa with the following codes and proportions out of the total continent area: code A is Tropical (31.0%); code B is Desert (57.2%), and code C is Temperate (11.8%). Peel et al. (2007) using Koppen's initial criteria determined for the hole earth the new area of the so called Tropical Savannah climate that amounts to 11.5%, ranging after the Hot Desert area (14.2%). The Tropical Savannah is also the most important among the three subdivisions of the tropical climate of Africa, bigger than the more known Sahel (BSh in the international classification). The area stretches from Senegal at the coast of the Atlantic Ocean to Ethiopia in the East, while expanding towards the South till Angola and Mozambique; in other words, it covers more than 20 countries. Tropical Savannah is classified with a temperature of the coldest month greater than 18 °C and a yearly rainfall range of 500-1500 mm mainly concentrated in half of the year, followed by a dry season of up to six months (Chesworth 2008). Since the African countries covered by Tropical Savannah, also called Tropical Grassland (Kusky 2005), have their economy based on agriculture, which in turn is largely dependent on the type of soils, these areas are of great interest.

2.2. Soil genesis

Many researchers consider that, under the combined effect of the weathering process and human activities, the Tropical Rainforest evolved to produce the Tropical Savannah

(Hillel 2004). The history and the characteristics of the soils in Tropical Savannah seem closely linked to those of Tropical Rainforest and Tropical Monsoon, with an important modification of the weathering conditions. In rainforest and monsoon climate zones, the driving forces of the weathering process are the high temperature and the almost permanent wet conditions, that lead to an intensive chemical alteration and leaching of the soil solutes (Alvarez-Benedí and Muñoz-Carpena 2005). This results in the dissolution and transformation of primary minerals like mica and feldspars and the loss of basic elements such as Ca, Mg, K, Na and silica while, in parallel, secondary minerals such as Kaolinite, ($Al_2Si_2O_5.H_2O$), Goethite (α-$HFeO_2$ or α-$Fe^{3+}O(OH)$) and the Gibbsite (α-$Al(OH)_3$) are produced.

Under certain conditions, another secondary mineral, the Lepidocrocite (γ-$Fe^{3+}O(OH)$ – red to reddish brown mineral formed of iron oxide hydroxide which often associates with Goethite – is also formed during a process called ferralitization (Hillel et al. 2004). Ferralitization leads to soils with low cations exchange capacity (CEC < 16 cmol/100g clay) because low activity clays such as Kaolinite dominate the products (El-Swaify and Emerson 1975). However ferralitization also leads to various oxide and hydroxide forms of iron and aluminium, hence generating the so called Ferrasols and Lixisols (International Rice Research Institute (IRRI) 1985). The Ferrasols result from a chemical and weathering process essentially driven by the long dry season such as met in Tropical Savannah. In these areas, the long dry season leads to the desiccation of the soils and therefore, the hydrated forms of iron and aluminium, in other words the hydroxides can lose their crystal water by dehydration, at least partially. As a consequence, Goethite is transformed to Hematite (Fe_2O_3), a very common iron oxide. This phenomenon explains why a striking difference exists between the humid tropical climates (*Af* and *Am*) yellow or yellowish brown soils – colour of the goethite (Nagano T et al. 1992) – and the red aspect (colour of the hematite) of the Tropical Savannah (*Aw*) soils (Hillel et al. 2004). These processes also explain why it is believed that Tropical Savannah was formerly occupied by Tropical Rainforest (*Af*) and later by Tropical Monsoon (*Am*) following the combined impact of long drought conditions and anthropic activities such as poor agricultural practices (Kusky 2005, Chesworth 2008).

2.3.   The prominence of iron and clay

The common occurrence of Hematite made of iron in the soils of the Tropical Savannah is justified by the origin of the soil parental material and the weathering process of earth crust. In terms of weight, the most common elements in earth crust are respectively oxygen (46.6%), silicon (27.7%), aluminium (8.1%), iron (5.0%), calcium (3.6%), sodium (2.8%), potassium (2.6%), magnesium (2.1%), and minor elements (1.5%) (Foth 1990). Though coming in 3rd position (Figure 2.1), iron is abundantly found in secondary minerals at an advanced stage of weathering. The weathering process attacks the parental rocks which are igneous rocks, sedimentary rocks, shale, and sandstones. The dominant minerals of these rocks are divided in 2 categories: i) primary minerals such as feldspars (also called albite), amphiboles, pyroxenes, quartz, micas, titanium minerals and apatite (that can also have a secondary origin), and ii) secondary minerals such as clays, iron oxides (goethite), carbonates and other minerals (Clarke 1924). The weathering process produces from parental minerals primary minerals that in turn weather further to yield secondary minerals which are less sensitive to weathering. Examples of these weather-resistant secondary minerals are not only quartz – which is known to be the mineral the most resistant to weathering and hence can be found accumulated in certain horizons in the soils – but also the iron oxides such as hematite (Fe_2O_3) so visible in by their vivid colours in Tropical Savannah.

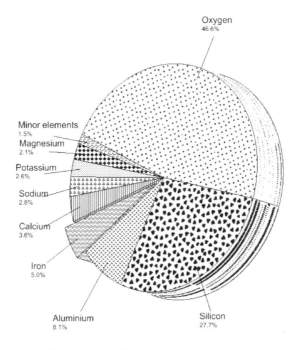

Figure 2.1: Proportions of basic chemical elements in earth crust

Data source: Foth (1990). The sections of the camembert represent the percentages in term of weigh for each basic element of the soil

Tropical Savannah soils are stable and essentially made of clay in humid areas which is generated at an advanced stage of the weathering process. The weathering is stimulated when the soil medium is acidic. In fact, the respiration of microorganisms and the roots of common plants (not aquatic plants like rice) generates in the soil carbon dioxide which reacts with water to yield the carbonic acid H_2CO_3. Therefore, the soil water becomes acid. Subsequently, this acidic medium stimulate the reaction of water with various minerals and is considered to be one of the most important factor triggering the weathering process (Jackson and Sherman 1953). A typical example is the reaction of feldspars (albite) with water and H^+ to produce kaolinite clay according to the reaction (Foth 1990):

$$2NaAlSi_3O_8 \left(albite\right) + 9H_2O + 2H^+ \rightarrow H_4Al_2Si2O_9 \left(Kaolinite\right)$$
$$+4H_4SiO_4 + 2Na^+ \tag{2.1}$$

This type of process the production of a secondary mineral from a primary mineral – tends to be the main cause of the persistent presence of clay particles in soils. In fact, the less than 2 μm particle sized secondary minerals produced this way like the kaolinite are called clay minerals, and they are resistant to further weathering. As a consequence, the primary material in the soil – here albite – disappears in time while secondary mineral like kaolinite accumulate (Foth 1990).

Tropical Savannah soils are old but with more silica – a basic element – than those of the Tropical Rainforest. The weathering from parental minerals to more weather-resistant products provides an indicator to assess the advancement of the process and hence to anticipate in what climatic areas a given type of soil can be found. The above transformation of albite has produced Kaolinite and silicic acid. Although extremely resistant, particularly in humid tropics (Murray 1994), Kaolinite (Eq.(2.1)) can be weathered and can disappear while the much more stable gibbsite is produced according to the equation:

$$H_4Al_2Si_2O_9 + 5H_2O \rightarrow 2Al(OH)_3(gibbsite) + 2H_4SiO_4.(silica) \qquad (2.2)$$

The two consecutive reactions (Eqs. (2.1) and (2.2)) can be schematized as follows:

$$albite \xrightarrow{-Si} kaolinite \xrightarrow{-Si} gibbsite. \qquad (2.3)$$

The monosilicic acid H_4SiO_4, produced from both the decompositions of the albite and the kaolinite, is an amorphous silica with high solubility and hence tends to be leached out of the soil profile, resulting into the loss of silicon *while alumina accumulates* as part of resistant mineral (Foth 1990, Murray 1994). On the basis of this observation, an indicator of weathering advancement expressed as the ratio silica/alumina (SiO_4/Al_2O_3) was proposed. This ratio drops for example from 3:1 for albite to 1:1 for kaolinite indicating the advancement in the weathering process. The silica/alumina ratio is usually greater (more sicila and less alumina present) in *Aw* Tropical Savannah where less leaching occurs than in Tropical Rainforest *Af* or Monsoon *Am* with important leaching. As a result, more weathering resistant minerals like gibbsite and hematite are more present under *Af* and *Am* tropical humid climates. When the medium is acidic with strong leaching, the only weathering products to be conserved are oxide clays with *alumina* and *iron* oxides and hydroxides (goethite, hematite...), conditions often present under humid tropics (Murray 1994) but also in valley bottoms of Tropical Savannah. Clays like kaolinite in an acidic environment such as often found in red coloured soils can lead – by protonation of the exposed hydroxyl OH of the clay sheet - to the formation of peds. The peds are extremely stable aggregates with a weather resistance equivalent to that of the quartz made of silica (Foth 1990). However, under hydromorphic conditions such as found in valley bottoms, the weathering leads to different clay minerals and colours.

2.4.   Differences between valley bottom and upland soils

While many authors agree about the necessity to clearly define valley bottoms, there are various attempts to attain a clear concept which should really distinguish key features of these specific soils. Various researchers characterize valley bottoms by considering three features: the temporary occurrence of flooding, the topography, and the hydromorphic conditions of the soils (MAHRH 1985, Zepenfeldt and Vlaar 1990, MAHRH 1999, FAO 2006). For example, Zepenfeldt and Vlaar (1990) proposed to define tropical valley bottoms as "flat or concave bottoms with temporary flow axes that are flooded during several days and where are found hydromorphic soils". Ministère de l'Agriculture de l'Hydraulique et des Ressources Halieutiques (MAHRH-BF) (1999) distinguished, at least for Burkina Faso, i) the flat valley bottoms as having both a transversal and longitudinal slope smaller than or equal to 5‰ and ii) the concave valley bottoms as having a transversal slope smaller than or equal to 5% and a longitudinal slope smaller than or equal to 1%. In addition, apparently not taking the criterion of "hydromorphic condition" of the soils as essential, Ministère de l'Agriculture de l'Hydraulique et des Ressources Halieutiques (MAHRH-BF) (1999) realised that the above definition could be so vague that it could include also alluvial plains and proposed to consider that valley bottoms transversal dimensions could not exceed 20-500 km. While the topography is important in the sense that it impacts on the flood extents in space and time, setting it to a particular value would not be required in the definition. Indeed, if hydromorphic conditions are well defined, conditions which spatial extent will not go beyond a certain limit because the soil saturated area is not infinite, the precisions about the slopes and the transversal dimension would not be necessary. Furthermore, setting a certain size or slope to a valley bottom is rather more a project-oriented condition than a characteristic to be included in the definition. The soil moisture conditions are the key feature for the definition of valley bottoms.

The hydromorphy of the valley bottom soils is greatly influenced by the duration of the flooding period that can be artificially extended for example by irrigation, and that makes a clear demarcation between them and those found in the uplands. Hydromorphic soils are also called gleysoils or redoximorphic soils (Hillel et al. 2004) and occupy about 2 million km² in the tropical areas of the earth. When well weathered minerals (goethite, hematite...) are located in areas subject to a relatively high but fluctuating groundwater table, oxidation and reduction occur in the 50-100 cm of the soil profile respectively during low and high groundwater table periods. A prolonged waterlogging causes saturation of the soil pores and, combined with the microbial oxygen consumption, will cause rapid depletion of molecular oxygen (O_2) – which, by nature, already has a low solubility in water (Murray 1994) – and anaerobic microbial proliferation. From aerobic, the soil switches to anaerobic. As a consequence, ferric iron F^{3+} in mineral compounds (reds or oranges for hematite, yellows and light browns for goethite) turns into ferrous iron Fe^{2+} (olive to dark blue hues) (Figure 2.2).

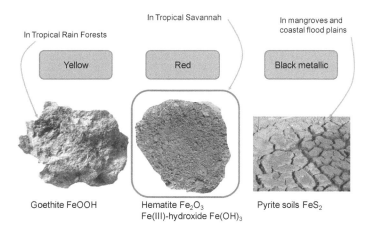

Figure 2.2: Soil types in West Africa

Hematite (red) is the most common soil found in Tropical Savannah, though its colour may turn to dark blue hues in due to oxidoreduction process in hydromorphic valley bottoms.

Because of its relative mobility, ferrous iron Fe^{2+} may be completely removed from the soil profile either by leaching or by crop roots intake and the soil turns to greyish hues (Chesworth 2008). Under frequent and prolonged flooding bringing more clay by sediment transport like in schemes with two flood-irrigation campaigns within a year, the greyish hue may be more permanent (Schlichting 1973). However, when the groundwater table drops in dry episodes and oxygen reintroduced, any iron present in the medium – and that is principally in the root channels and soil fissures – is reconverted in a ferric status Fe^{3+} and the soil colour locally turns back to red and yellow hues. Therefore, redoximorphic or gleysoils are recognizable by their greyish hue (absence of iron F^{3+}) with the presence in the fissures of yellow and red hues (goethite, hematite). The Food and Agriculture Organisation (FAO) goes further in the definition of gleysoils mentioning that it should be observable over 50 cm (FAO 1993). This redox process explains why Tropical Savannah valley bottom soils, with their greyish colours, do not look like the more general group of red soils (hematite) found in the uplands of Tropical Savannah.

2.5. Iron toxicity

Iron toxicity is primarily caused by the excessive uptake of ferrous iron Fe^{2+} by the rice roots in the soil solution. This uptake can lead to the formation of oxygen radicals, which are phytotoxic and responsible for protein degradation and peroxidation of membrane lipids. The symptoms of iron toxicity (often similar to those of Zn deficiency) are as follows. Tiny brown spots develop on the leaf tip and spread towards the leaf base. Symptoms first appear on older leaves. Under severe Fe toxicity the whole leaf surface is affected. Leaf bronzing occurs also in K deficient rice plants which are unable to maintain sufficient root oxidation power (Achim and Thomas 2000). Iron toxicity alters the root structure, the crop development, and leads to sterility. In 2001, a preliminary survey by the West Africa Rice Development Association (WARDA) estimated that as much as 60% of the lowland rice in West and Central Africa may be under threat of iron toxicity (Figure 2.3).

Figure 2.3: Iron toxicity spread in West Africa

(A) Partial iron toxicity map in West Africa. (B) Case of ferrouss iron resurgence in a valley bottom irrigated rice scheme. Source: WARDA (2002)

The average yield loss due to iron toxicity is tremendous, ranging from 12 to 100% (West Africa Rice Development Association (WARDA) 2002). Therefore the problem is crucial, particularly in the Tropical Savannah where most of the valley bottom rice production systems are located. In valley bottom fields, the existing poor drainage maintains quasi-permanent anaerobic conditions favourable to the development of sulphate and iron reducing bacteria (SRB and IRB). Several researches pointed to these bacteria as the main catalysers to chemical reactions leading to the reduction of iron and sulphate into toxic products for rice (Jacq 1989, Ouattara 1992, Dianou 2005). In valley bottoms systems, the problem is aggravated by the anaerobic baseflow and groundwater flow carrying the ferrous ions Fe^{2+} from the upland areas and the slopes to the valley bottom where rice and other crops are grown. In many rice irrigation systems located in such areas, the resurgence of groundwater solutions containing ferrous iron can easily be observed in farm plots (Figure 2.4).

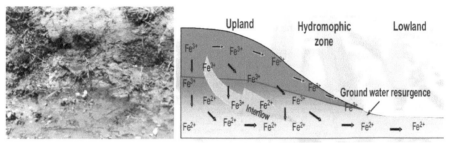

Figure 2.4: Close view on iron reduction

Groundwater resurgence zone at edge of valley bottom field where the slope and stationary water meet. Note reddish ferric iron on soil surface becoming paler under the process of reduction.
Source: WARDA (2002), (2006)

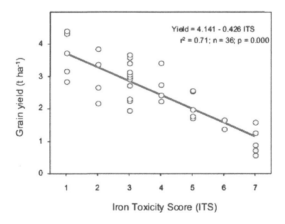

Figure 2.5: Correlation between iron toxicity score and grain yield for 12 cultivars (1998)

Source: WARDA (2002)

For several decades, WARDA has been leading research against iron toxicity using various approaches. One of these ways is the development of new iron-toxicity tolerant cultivars. It is expected that with modern breeding technologies, the incorporation of tolerant genes from traditional cultivars to a strong crop would render more tolerant rice in areas with iron toxicity issues. There were encouraging results on tests of resistance with experiments on 12 cultivars which showed an average grain yield drop of 0.5 tons/ha for every increase of 1 on the Iron-Toxicity Score scale (ITS) (Figure 2.5). NERICA for uplands varieties developed from *Oryza Glaberrima* (the African rice) – with good resistance to drought and aluminium – are currently introduced in several African countries. However, WARDA recognized in 2006 that "The new Lowland [valley bottom] NERICA lines are yet to be widely tested under iron toxicity stressed environments". Meanwhile, different methods, summarized on Figure 2.6 were proposed by WARDA to alleviate the effect of iron toxicity on rice production soils of Africa. For example, 'resistant cultivars' are suggested for valley bottom flooded rice systems. In addition, cultural practices are mentioned. However, one can notice the absence of suggestion about using subsurface drainage to oppose excessive of ferrous iron Fe^{2+} uptake by the rice.

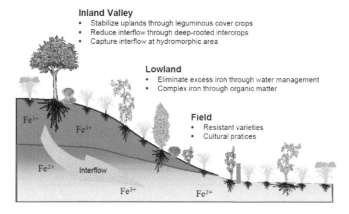

Figure 2.6: Toposequence diagram showing options for reducing iron movement and toxicity

Source:WARDA (2002)

Rice production in Asia, Africa, and South America is widely affected by iron. Ferric iron Fe^{3+} is naturally present in aerobic soils. When anaerobic conditions occur, such as long flooding, a large part of this iron can be reduced to ferrous iron Fe^{2+}. Hence, iron toxicity may appear in plants due to excessive Fe^{2+} absorption by roots (Ponnamperuma 1972, Sahravat KL 2004, Majerus et al. 2007). Wetland rice can – as many other aquatic plants – survive in flooded soils because of its aerenchymatic air transport ability from the shoot to the roots where air is released into the rhizosphere. However, this oxidative protection – in certain physicochemical soil conditions – may fail resulting in ferrous iron or sulphide toxicity. That happens for example when facultative anaerobic iron or sulphate reducing bacteria rapidly deplete the oxygen released by rice and start to use ferric iron Fe^{3+} as electron acceptor (instead of O_2), hence producing ferrous iron Fe^{2+} (Scala et al. 2006). The accumulation of black ferrous sulphide in the rhizosphere and on the roots is reported in all cases where iron or sulphide toxicities are involved. But this accumulation was also observed in the root zone of younger growing plants in alluvial deposits and mangroves in Senegal (Jacq et al. 1988).

2.6.   Sulphate-reducing bacteria in soils

2.6.1. The flooded rice ecosystem

Rice ecosystem is a composite medium which equilibrium is constantly disrupted by cultural practices – like irrigation, fertilization, seeding, etc. – and natural phenomena (such as rainfall, solar radiation, temperature, etc.). As Par suggested (1991) the rice ecosystem can be viewed and composed of five sub-systems: 1) the water layer above soil surface, 2) the oxidized surface soil, 3) the reduced soil, 4) the subsoil, 5) the rice crop and its phyllosphere and rhizosphere (Figure 2.7). In each subsystem, there are levels in which the biotic communities grow separately. The interactions among the subsystems are a key factor in understanding the energy and matter transfers in a rice field. Every subsystem possesses its own chemical and biochemical properties so that specific algae, mushroom and bacteria populations grow in it (Ponnamperuma 1972). The activities of the algae and bacteria are highly correlated to nutrients availability such as the sources of carbon, nitrogen, phosphate, phosphorus and other energy sources. On the other hand, these activities depend on physicochemical conditions in the medium such as the oxidoreduction potential ORP, the temperature, the pH, the salinity and so on (Ouattara 1992).

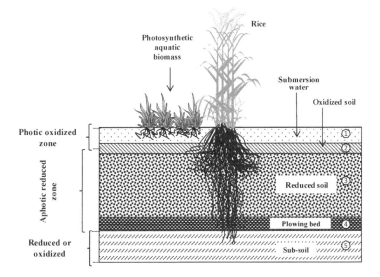

Figure 2.7: Five subsystems of flooded rice ecosystem

Source: adapted from Par (1991)

The decomposition of the straw, the photosynthetic biomass (algae and weeds) and the root exudates (Figure 2.7) provides the energy and carbon (Figure 2.8) required for the activities of the microorganisms (Jacq 1989). When the required conditions to trigger these microbial activities occur, they follow a sequential order, from aerobic bacteria to methanogenic bacteria according to the quantity and the type of electron acceptors available in the medium. However, the activities can take place simultaneously (Ouattara 1992). Ferric iron and sulphate are electron acceptors (Figure 2.8) the products of which are toxic for rice crop. Iron intoxication seems due to the penetration of ferrous ions in the plant roots or to the accumulation of iron sulphide FeS on the roots (Prade et al. 1990, Jacq and Ottow 1991).

The sulphate reduction is a frequent phenomenon observed in rice fields. It is estimated to be generating up to 95-97% of the total sulphide products in these biotopes. The sulphate and iron reducing bacteria (SRB and IRB) are often present in flooded rice fields. The SRB and IRB are localized in the reduced horizons but also in anoxic micro sites of more oxygenated horizons of the soil (Figure 2.7). The number of these bacteria varies between $10^1 - 10^3$ bact./g of soil in non-flooded rice field while this number reaches 10^4-10^9 bact./g of soil in flooded rice fields (Moormann and Breemen 1978, Watanabe and Furusaka 1980, Jacq 1989, Andreisse and Fresco 1991).

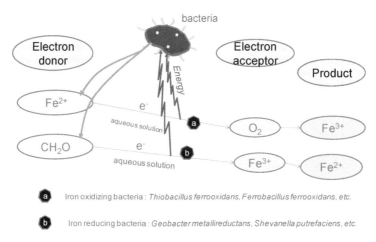

Figure 2.8: Bacteria roles in iron oxidoreduction

Bacteria use energy released during the transfer of electrons
from Fe^{2+} to O_2 or from organic matter CH_2O to Fe^{3+}.

2.6.2. Iron and sulphate reducing bacteria

During the dissimilatory reduction process of ferric iron Fe^{3+}, the energy liberated is used by the bacteria for growth (Figure 2.8). This process is made possible by the presence in the medium of an electron donor and leads to the production and accumulation of ferrous iron Fe^{2+}. The process can be considered as an oxidation in which oxygen is replaced by another electron acceptor, namely Fe^{3+}. It seems that ferric iron is the most abundant electron acceptor in the nature (Lovly 1987).

The dissimilatory reduction of iron is not well investigated most probably because very few microorganisms use it as an obligated way for energy capturing. However, a great number of aerobic, anaerobic (strict or facultative) microorganisms can reduce iron (Ouattara 1992). Some rare microorganisms such as *Geobacter metallireductans*, *Desulfovibrio desulfuricans*, and *Defulfotomaculum nigrificans* use the dissimilatory reduction of iron Fe^{3+} as a source of energy conservation. Other bacteria like *Shewanella putrefaciens* couple iron reduction and the oxidation of hydrogen or the formic acid (H-COOH). *Geobacter metallireductans* can realise a full oxidation of the substrates they use, such as the acetate, propionate, butyrate, ethanol, benzoate, etc.. Interestingly, in all known cases, the iron-reducing bacteria can also use the nitrates (NO_3^- compounds) as electron acceptor (Ouattara 1992).

Ouattara (1992) wrote that the molecular mechanisms of iron reduction were not elucidated yet. Nevertheless, he provided the identity of 31 iron-reducing bacteria as the result of the research of various researchers whose references were also provided (Table 2.1). But in his otherwise very useful review of Microbiology of Flooded Rice, Liesack and al. (2000) inadequately wrote - after mentioning nitrate reducer, iron Fe(III) reducers and sulphate reducers - that 'little' was 'known about the identities of the major microbial players [...]'. In fact, it seems that very few among the authors mentioned by Ouattara in Table 2.1 are taken into account in the review of Liesack. Furthermore, neither Ouattara, nor Jacq are mentioned.

Studying the effect of fertilizer deficiency on the growth of iron-reducing and sulphate-reducing bacteria in Senegalese flooded rice fields, , Jacq et al. (1988) arrive to the following conclusions:

- nutrient deficiency impact is higher on iron-reducing bacteria (IRB) than on sulphate-reducing bacteria (SRB);
- in poorly fertilized soils (no application of P, K, Ca and Zn) and at vegetative stage: the rice roots are permeable at the vegetative stage and high degree of exudation of the roots make the plant vulnerable to ferrous intoxication;
- in well fertilized soils (appropriate application of P, K, Ca and Zn) but only at flowering stage: the roots undergo decaying and senescence with a low oxidative power leading to an increasing anaerobic medium in the rhizosphere. As a consequence, the ferrous intoxication induced by the bacteria occurs.

Therefore, fertilization seems to have a higher impact in poorly fertilized than in well fertilized soils, though both are sensitive to some extent.

2.6.3. The bacteria against nematodes

Sulphate-reducing bacteria, though known being catalysers of products toxic for rice, were searched for their role in controlling another important rice enemy: the rice nematode *Hirschmanniella oryzae*. In a study carried out using over ground microplots lined with concrete, Jacq and Fortuner (1979) proposed using the SRB to destroy the nematode in rice fields. A series of three experiments – the first of which was a pilot one – were conducted on 1 m x 1 m x 1 m microplots of ORSTOM[2] Laboratory in Dakar, Senegal. The soils were brought from a rice field near Dakar and autoclaved before introduction in the microplots. This was to insure an initial condition free from both bacteria and nematodes. The following sets of soil variables[3] were measured: 1) the total sulfur, 2) the organic carbon, 3) the clay proportion, and the *pH* for acidity (Table 2.2).

The variety of rice used was I.R.8. The team noticed also that nine microplots had cracks in their walls and thus were subject to a (non-measured) seepage. The pilot experiment and the experiment N1 used 15 microplots (nine with cracks). The experiment N2 used 20 other microplots (Jacq and Fortuner 1979).

---

[2] Office de la Recherche Scientifique et Technique Outre-Mer. Currently IRD
[3] Which can be considered as the confounding variables.

Table 2.1: Iron- reducing bacteria

Organisms	References
Actinomucor repens	Ottow and Von Kopotek, 1969
Areobacter areorogenes	Ottow, 1970
Aerobacter sp.	Bromfleld, 1954a; Ottow, 1970; Troshanov 1968
Alternaria tenuis	Ottow and Von Kopotek, 1969
Alteromonas putrefaciens	Semple and Westlake, 1987
Bacillus cereus	Ottow, 1970
Bacillus circulans	Bromfield, 1954a; Ottow, 1970; Troshanov 1968
Bacillus mesentericus	Troshanov, 1968
Bacillus polymyxa	Roberts, 1947; Hamman and Ottow, 1974; Bromfield,1954a
Bacullus pumilus	Ottow, 1970
Bacillus sp.	De Castro and Ehrlich, 1970
Bacillus subtilis	Ottow and Glathe, 1971
Bacteroides hypermegas	Jones et al., 1984
Clostridium butyricum	Hamman and Ottow, 1974
Clostridium polymyxa	Troshanov, 1968
Clostridium saccharobutyricum	Hamman and Ottow, 1974
Clostridium sporogenes	Starkey and Halvorsen, 1927
Desulfovibrio desulfuricans	Jones et al., 1984
Desulfotomaculum nigrificans	Jones et al., 1984
Escherichia coli	Starkey and Halvorsen, 1927
Fusarium oxysporum	Gunner and Alexander, 1964
Fusarium solani	Otto and Von Klopotek, 1969
Geobacter metallireductans	Lovley and Phillips, 1988; Lovley and Lonergan, 1990
Paracolobactrum sp.	Bromfield, 1954a
Pseudomonas aeruginosa	Ottow, 1970
Pseudomonas denitrificans	Jones et al., 1984
Pseudomonas liquefaciens	Troshanov, 1968
Pseudomonas sp.	Ottow and Glathe, 1971; Balashova et Zavarzin, 1980
Serratia marcescens	Ottow, 1970
Shewanella putrefaciens	Obuekwe et al., 1981; Myers and Nealson, 1990
Sulfolobus acidocaldarius	Brock and Gustafson, 1976
Vibrio sp.	Jones et al., 1984

Source: adapted from Ouattara (1992)

Table 2.2: Physico-chemical properties of the soils

Observations on rice	Total sulphur (%,w/w)	Organic carbon (%,w/w)	Clay (%,w/w)	pH
Experiment 1	0.4-1.4	1.14-1.34	44-48	7.8
Experiment 2	0.2-1.1	1.60-1.80	43-45	8.2

Source: Jacq and Fortuner (1979)

The measurements were designed with the aim to establish a cause-effect relationship. The alternative hypothesis can be formulated as follows: "soluble sulphides are able to limit nematode populations if the activity of sulphate-reducing bacteria (SRB) is sufficiently high". In other words, there were two hypotheses the first of which was: "If the SRB population is increased, then the quantity of soluble sulphide will increase"; and "if the quantity of soluble sulphide increases, then the amount of nematodes will drop". To test these two hypotheses, the authors designed the experiment described above. The four independent or *factor* variables (Mason et al. 2003) measured were: 1) the rice growing time (days), 2) the microplots soil conditions (aerobic with cracks in the covering or anaerobic), 3) the soil moisture duration (days of waterlogging situation or non-waterlogged situation (called "dry")), and 4) the amount of fertilizer used (urea, SCU or sulphur).

The "effects" or *response* variables were the following ones: 1) the number of sulphate-reducing bacteria (Nbr/kg of dry soil), 2) the total sulphide (in 10^{-6} $S^=$), 3) the soluble sulphide production (in STU or Sulphide Time Units; 1 STU = 1 nematode placed for 1 day in a solution of 1 ppm sulphide concentration), 4) the growth of nematodes *H. oryzae* (in percentage of the initial population of nematodes in the microplots), and 5) the paddy rice yield (in grams of rice harvested by plant).

The results of the second experiment shown of Figure 2.9 provide very useful insight. Though the authors did not provide any statistical analysis about the confidence interval, the trend in graph (A) was similar for the 3 experiments and shows the increase of SRB in time. Furthermore, it can be noticed that their number is lower in the control plots (curve a), though these curves are very close at the end of the trial period. The microplots conditions b1 and b2 differ in the flooding management: the plots were not flooded (no water layer on top of the soil) from day 36 to day 55 (20 days) in b1, while the non-flooding period was day 36 to day 69 (34 days) for b2. On graph B (Figure 2.9), the soluble toxic sulphide increased in time and this change was assumed to be the results of the reducing activities of the SRB. The curve a is absent because no soluble sulphide was produced in the control plots. The researcher did not link clearly the water management variables with the rice growing stages. The 2 curves b1 and b2 are very close on Figure 2.9 (B) and no significant conclusion can be drawn. However, the effect of the increase in toxic sulphide concentration on the nematode populations is slightly visible during the rice cycle as shown in Figure 2.10. Again no statistical analysis was available about the significance of the measurements.

According to Jacq and Fortuner (1979), the peak around day 35 is explained by the existence of a "cool" weather with temperatures as low as 15°C during the night and 21 °C during the day. This "cool" weather had reduced the activity of the SRB, resulting in the peak of nematode population. It seems that the temperature in the soil of the microplots was not measured as a factor. It might be postulated that change in SRB population and the nematodes are both mainly induced by the temperature change.

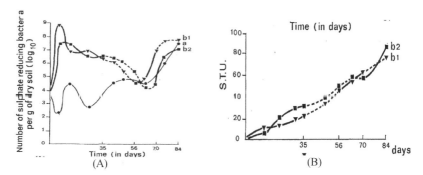

Figure 2.9: Exp N2 - Number of SRB (A) and soluble sulphide toxicity- STU (B) in 20 microplots

a = control plots (dry soil with crack wall covering); b1, = treated plots, waterlogged during days: 0-35 then 56-70; b2, = treated plots, waterlogged during days: 0-35 then 70-55. Source: Jacq and Fortuner (1979).

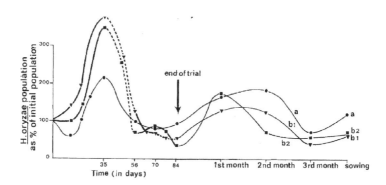

Figure 2.10: Exp N2 - Nematode Hirschmanniella o. population change during the rice cycle

a = control plots (dry soil with crack wall covering); b1 = treated plots, waterlogged during days: 0-35 then 56-70; b2 = treated plots, waterlogged during days: 0-35 then 70-55. Source: Jacq and Fortuner (1979).

Several meaningful lessons can be drawn from the experiments. An important conclusion formulated by the researchers is that the nematode population between the beginning and the end of the trial period remains the same: it was controlled by the SRB activity. But the authors reported that rice was damaged in the pilot experiment with high SRB activity, up to 15-35% of plant mortality. Another result, though not aimed, is that in microplots with a "subsurface drainage" due to the existence of cracks in the lining, the SRB activity was reduced, due to more air and oxygen penetration in the rootzone according to the authors. Also, the experiments conducted to higher rice yields, attributed to the reduction of the nematodes, reduction in turn caused by the production of more soluble sulphide. But the soluble sulphide is also fatal for rice itself. Therefore, the applicability of this method remained difficult. Furthermore, the authors explained that there was a significant build-up of nematodes after removal of water in the treated plots. Finally, one can observe that the experimentations do not describe how many potential lurking variables - such as the soil temperature, the oxygen content of the water, the source of water, and the discharge per seepage through the cracks - were taken into account. Finally, no significance was provided.

2.6.4. The reduction processes

Different authors studied the reduction processes in flooded rice using the standard electrode potential redox. The main equations are summarized in Table 2.3. All reactions in this table are catalyzed by bacteria. The positive numbers of the standard electrode potential at 50% of reduction and with a *pH* 7 are related to reactions producing energy (last column). These reactions can thus take place spontaneously. The last three processes are energy absorbers. An external source of energy is required for them to occur. This means the electron acceptors mentioned in the 4th column are increasingly "reluctant" and when it comes to $Fe(OH)_3$, an external energy (Table 2.3) is required to displace the equilibrium toward the reduction (Jacq et al. 1988). There is only a slightly reduction of Fe^{2+} -oxides and sulphates in a sterile medium. However, ferric iron reducers and sulphate-reducing bacteria can use these compounds as electron acceptors during energy conservative metabolisms (Figure 2.8). The energy conservation by the bacteria mineralizing organic matters is more efficient than the fermentative process (Berthelin and Kogblevi 1974, Berthelin J and Boymond D 1978, Ottow et al. 1982).

In the rice soil, with the increasing availability of easily decomposable organic matter, the demand of electron acceptors increases also. The consequence is the start of the sequential reduction of available electron acceptors (Table 2.3). The reduction of nitrates and Mn^{4+} -oxides occurs very soon after the soil flooding. Therefore, the main redox-buffering capacity is provided by the Fe^{3+} -reducible compounds and sulphates (SO_4^{2-}) that appear then to be the prevalent electrons acceptors (Inubishi K. H. 1984). *However, the sulphate reduction will not occur in bulk soil as far as Fe^{3+} -reducible compounds are available – which is the case in hematite soils – because less energy is required for this reaction.* The formation of the toxic hydrogen sulphide (H_2S) from sulphates takes place only in those soils compartments where the ferric oxide compounds are exhausted. The decomposition process from complex organic compounds to mineral toxic compounds such as iron sulphide (FeS) and hydrogen sulphide (H_2S) is illustrated on Figure 2.11.

Table 2.3: Process sequences, redox and theoretical half reactions at pH 7 in flooded soils

Processes	Measured redox potential *ORP* (mV) during transformations*	rH level	Redox systems involved	E'$_0$(mV)**
Respiration	> +400	>26	$O_2 + 4H^+ + 4e^- \leftrightarrow 2H_2O$	+814
NO_3^- respiration	+500 to +203	29-19	$2NO_3^- + 12H^+ \leftrightarrow N_2 + 6H_2O$	+741
Formation of Mn^{2+}	+400 to +203	26-19	$MnO_2 + 4H^+ + e^- \leftrightarrow Mn^{2+} + 2H_2O$	+410
Fe^{2+} production	+400 to +183	26-18	$Fe(OH)_3 + 3H^+ + e^- \leftrightarrow Fe^{2+} + 3H_2O$	-155
S^{2-} production	+100 to -200	16-5	$SO_4^{2-} + 10H^+ + 8e^- \leftrightarrow H_2S + 4H_2O$	-214
CH_4 production	-150 to -280	7-2	$CO_2 + 8H^+ + 8e^- \leftrightarrow CH_4 + 2H_2O$	-244

*Redox potential (*ORP*) are comparable only if corrected for pH

**E'0 = standard electrode potential at 50% of reduction at a pH 7.0

Sources: (Patrick 1960, Yoshida et al. 1976, Ottow and Fabig 1985)

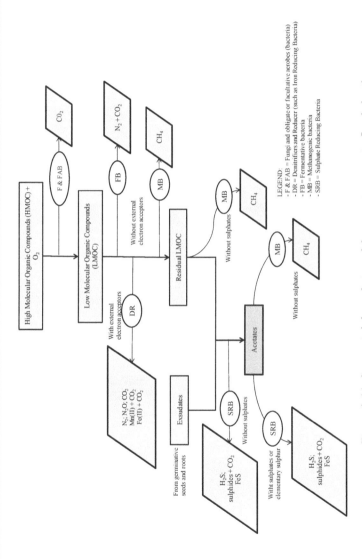

Figure 2.11: Sequential degradation of the organic matter in flooded soils

Source: Freney JR et al. (1982); Jacq et al. (1988)

In an interesting way, Liesack et al (2000) describe the flooded rice soil as medium in which complex microbiological and chemical processes take place. The medium can be viewed as made of 3 compartments characterized by different physico-chemical conditions: 1) the oxic surface soil, 2) the anoxic bulk soil, and 3) the rhizosphere (Figure 2.12). The oxic soil comprises not only the first centimetres of soil surface but also a thin layer around the plant roots where oxygen is released through aerenchymatic air transport. Therefore, there are two types of oxic/anoxic interfaces: one between the oxygenated soil surface and the second at the border of the oxygen "layer"covering the roots. After flooding, oxygen is rapidly depleted, consumed by the respiration of aerobic bacteria and oxidative chemical reactions. Thus, in anoxic zone, alternative electron acceptors are used. Nitrates ($NO3-$ compounds) are the first electron acceptors to be used after oxygen depletion, followed by Mn^{4+}, Fe^{3+} $SO_4^{2-}$and CO_2 (Table 2.3 and Figure 2.12). At the oxic/anoxic interfaces, a partial regeneration of alternative electron acceptors occurs. The final reduction stage is the production of methane CH_4 for the carbon dioxide CO_2.

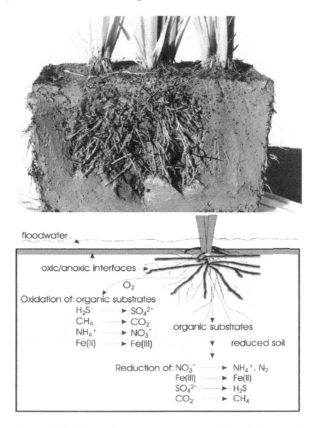

Figure 2.12: The different compartments of flooded rice soil

Source: Liesack et al. (2000)

The presence of oxygen can serve as a basis to build a strategy against iron toxicity. As said above, in the rice paddies not only the surface soil but also the thin coat around the roots contains oxygen released by the aerenchymatic air transport of the rice and other aquatic plants. Oxidation may therefore occurs: oxidation of methane (CH_4) (DeBont et al. 1978, Holzapfel-Pschorn et al. 1985, Gerard and Chanton 1993, King 1994, Gilbert and Frenzel 1995), ammonium (NH4[+]), sulphide (H_2S) (Reddy KR et al. 1989, Wind and Conrad 1995, Arth et al. 1998), ferrous iron (Fe^{2+}) (Ahmad and Nye 1990, Begg CBM et al. 1994, Kirk GJD 1995). Therefore, oxygen seems to be a key factor controlling the gradient of other electron acceptors in the soil (Liesack et al. 2000). The renewal of water in the root zone by subsurface drainage can be reasonably envisaged as a mean to stop or alleviate the reduction processes leading to iron and sulphide toxicities. It is assumed that the drainage water will remove certain populations of bacteria and that the new irrigation water will carry more oxygen into the root zone.

2.7.   Subsurface drainage

In order to assess the impact of subsurface drainage on crop production and soil quality in valley bottom acid sulphate soils in south-west India (Kuttanad in Kerala State), Mathew et al. (2001) undertook a field research with two experiments (N1 and N2). One was oriented toward crop production parameters, the other toward soil quality. The set up for the first experiment is summarized in Table 2.4. As drain pipes, clay tiles with 100 mm/125 mm of internal/external diameters were used (Figure 2.13). The pipes were placed into the soil 87 cm below surface level in a total farming area of 2.5 ha. Two systems were designed: one with a drain spacing of 15 m and the other of 30 m. Pipes at the two edges and the transitional area between the two system were excluded to preserve a certain flow symmetry. These drain pipes were called 'buffer lines'. After two years of measurements, the researchers obtained in experiment N1 several interesting results. For example, the plant height at maturity, the number of plants per square meter or the rice grain yield was found superior in the subsurface drained areas than in the control.

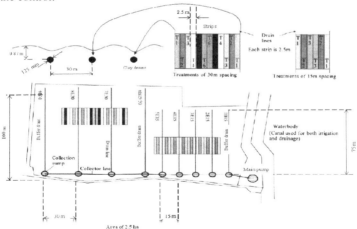

Figure 2.13: Kuttanad subsurface drainage experiment layout

Source: adapted from Mathew et al. (2001)

Subsurface drainage of valley bottom in Tropical Savannah

Table 2.4: Experiment N1 set-up variables of the Kuttanad field research

Variables	Topography	Sub-surface drain system	Soil	Water	Plant
Initial parameters (confounding variables)	$Ilat$ (%) = 0.20 [drain lat slope]; $Lstrip$ = 2.5 m [sampling strip width] $Jlong$ (%)= longitudinal slope of drains [not provided] $Dist(P)$ = point P in the north or south of a strip $Twidth$ = 2.5 m [width of the soil sampling strips] 8 replication sets with each set containing 6 soil sampling strips in the 15m drain spacing zone/ 4 replication sets with each set containing and 6 soil sampling strips in the 30 m drain spacing zone 8 control sets taken in 'area with no influence of the drains' for the 15 m drain spacing zone [location not provided]/4 control sets taken in 'area with no influence of the drains' for the 30 m drain spacing zone [location not provided]	$\Phi drain,ext$=125 mm [drain external diameter]; $\Phi drain,int$=100 mm [drain internal diam]; $Dpip$ = 87 cm [drain depth]; $Lpip$ =75 m [drain length] $SPdrain$ = 15 m for 5 drains and $SPdrain$=15 m for 4 drains [drain spacing] Out of the 9 drains, 3 are 'buffer drains' (not studied because in dissymmetrical zone)	Surface of 2.5 ha belonging to a group of 4 farmers Acid sulphate soil [analysis not provided] $Dimper_v$ (m)= Impervious layer of soil [not provided]	$Dtime$ = 8 h (drainage time) Farmers 'normal' irrigation	Rice variety = Pattambi-39 Fertiliser used [not specified]
Independent variable (manipulated and measured)				Farmers irrigation water [not manipulated] Farmers drainage [not manipulated]	
Dependent variables (measured)				$pHdrain$ = pH of drained water; $ECdrai$ = electrical conductiv of drained water	$Hmatur$ = crop height at maturity; Npl / m^2 = number of plants per 1m²; $Npan/m^2$= number of panicle per 1m²; Ngrain/pan = number of grain per panicle; $Ygrain$ (t/ha)= grain yield; $W100grain$= weight of 100 grains; $Ystraw$ (t/ha)= straw yield;

Note: In the experiment N2, the independent variables were the same as in experiment N1. However the dependant variables were different. During 2 years, were measured: pH, EC, Fe, $SO4^=$, Ca, Mg, Na, K and Cl concentrations for soil samples. The study area is the same and controls with 'no drainage' were used [no location specification]

However, many other results were not found making significant differences. For the weight of 100 grains and several other variables such as the number of panicles/m², the straw yield, the chaff, the control results were equal or superior to those of the treatments. As regards with the experiment N2 dealing with soil/water physical and chemical properties, the researchers report that there were no control plots (see Table 2.4 Note for the dependent variables measured).

Though it is reasonable to presume that subsurface drainage by renewing soil oxygen content will have positive effects on rice crop, the design of the experiments of Kuttanad included several potential sources of bias. First, conducting such an experiment to assess the benefits of subsurface drainage on flooded rice production using a group-managed farm is too difficult to follow up in term of variables. Though many of the initial parameters were well defined, several others were not. For example (Table 2.4), no details were provided about the control plots: where were they located? If they were in the same treatment zone, how could the drains effect be prevented on those controls? If they were located outside the treatment zone, how can one be sure that the soils, the farmers' practices in term of irrigation or land management were identical to those of the treatment? The controls are of extreme importance in environmental sciences since they help to eliminate the effect of the lurking variables, which are generally in great number and difficult to discern (Table 2.4). Without a proper control design, the internal validity of the experiments fails (Shuttleworth 2008) and the results, such in the case of the Kuttanad, cannot be said supporting the alternative hypothesis namely: 'subsurface drainage improves flooded rice production'.

In the same line, the design does not provide any detail about the fertilizers and the chemical use of the farmers in the different plots. It is only said that they proceed as usual. Anything is reported about the longitudinal slopes of the drains or the depth of the impervious layer in the soils. These factors may have contributed to the arguable results. However, the difficulties of this study support the proposal to make the research on drainage impact on rice yield on microplots, where the lurking variables can be much easier to neutralize through the use of controls.

2.8.   Concluding remarks and knowledge gaps

Iron toxicity is a problem of serious extend in valley bottom systems of Africa

The surveys of WARDA and others researchers investigation show that iron toxicity is of serious concern in valley bottom or lowland rice production systems. WARDA in 2001 announce a figure of 60% of the valley bottom rice in West and Central Africa as being under the threat of iron toxicity. The yield loss can reach 100% in some irrigated fields in Senegal; (Jacq et al. 1988, West Africa Rice Development Association (WARDA) 2002)

Subsurface drainage can have positive effect on SRB / Field experiment related to SRB are difficult due to many lurking variables

Though designed to prove that sulphate-reducing bacteria can help to improve rice yield (by killing nematodes), the microplots experiments of Jacq and Fortuner (1979) show that SRB activities are damaging for the crop. The accidental subsurface drainage through the wall cracks of the microplots tends to support the hypothesis that subsurface drainage would be beneficial for the crop production. The difficulties encountered by the researchers show the necessity of defining clearly the dependant and independent variables of such an experiment. They stress particularly the importance of well designing the controls in the experiment to neutralize the effect of the confounding variables.

The impact of SRB and IRB on rice yield is not scientifically ascertained

The experiments undertaken for example by Jacq and Fortuner (1979), Mathew et al. (2001) tend to support that the neutralization of SRB and IRB activities will lead to higher rice yield. However, the difficulties in the design of experiment and the ambiguity of certain results show that there is still a need to ascertain the real impact of SRB and IRB population and effect on the diminution of rice yield.

Oxygen seem to be a key factor in the soil to control SRB and iron toxicity

As it can be seen in the analysis made by Liesac et al. (2000), oxygen seemed to be a key factor controlling the gradient of other electron acceptors in the soil. Studying their literature review, it seems reasonable to assume that the renewal of water in the root zone by subsurface drainage can be a mean to stop or alleviate the reduction processes leading to iron and sulphate toxicities.

3. MATERIAL AND METHODS

3.1. The two study areas

Burkina Faso, one of the seven Sahelian[4] countries, holds distinct climatic zones which deeply influence the type of soils, water management and agricultural activities. Several regional and local subdivisions exist about the climate in Burkina Faso. One of them classifies three zones (Ministère de l'Agriculture de l'Hydraulique et des Ressources Halieutiques (MAHRH-BF) 1999): Sahelian (350-650 mm), Sudano-Sahelian (650-900mm) and Sudanian (900-1200 mm). However, according to the international climate classification of Köppen (Koppen 1936, Peel et al. 2007), there are mainly two zones in Burkina Faso, which are the Arid Steppe with an annual rainfall smaller than 500 mm, and the Tropical Savannah with a rainfall between 500 mm and 1200 mm (Figure 3.1). The soils in the environment are essentially made of hematite (Fe_2O_3) (Spaargaren and Deckers 2004) .The western and south western regions – in contrast with the rest of the country – are rather well-watered areas, though there still exists two clearly separated rainy season (May to October) and dry season (November to April). Therefore, it is not surprising that irrigation is practiced in those regions. In fact, several reservoirs – the first reason of the existence of which was to create water points for the populations and the livestock – were implemented and are now providing supplemental irrigation during the rainy season and full irrigation in the dry one. The reasons of the construction of these small reservoirs – often less than 1 million m^3 – have changed in time. Several drought events (early 1970s and 1980s) have revealed the high vulnerability of the local population and their livestock to rainfall. The reservoirs contribute to attain food security and to alleviate poverty.

3.2. The site of Tiefora

3.2.1. Geographic location

Tiefora is a 16 ha-irrigation scheme using a small dam and a valley bottom to provide a very useful support to rice growing. The coordinates of the site determined by GPS are: 4°33'13.19' longitude West and 10°37'33.56' latitude North (Figure 3.1). The village of Tiefora, near the dam of the same name, is located at a distance of 25 km from Banfora which is the county town of the Comoé province in Burkina Faso. The dam, which coordinates are 4°33'18.50' longitude West and 10°37'18.00' latitude North, is located on the right side of the national road N11 connecting Banfora to Sideradougou. The valley bottom rice field, object of this research work, was constructed 800 m downstream the dam. One can access the site by the road from the central market to the

---

[4] Sahel is the semi-desert transition zone between the Sahara desert and the humid climatic zone of Africa. It stretches across the seven following countries: Mauritania, Senegal, Mali, Burkina Faso, Niger, Chad, and Sudan.

upstream side of the valley, passing in front of the prefecture (Office National des Barrages et des Aménagements Hydroagricoles (ONBAH) 1987).

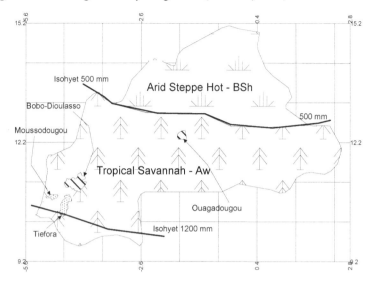

Figure 3.1: Climatic zones and location of Tiefora and Moussodougou

The three climatic areas distribution from isohyets based on rainfall data from 1960's to 1990's

3.2.2. Population

Although it is formed by various villages and ethnic groups, the population of Tiefora gets its revenue essentially from agriculture. The town of Tiefora, with 4700 inhabitants (Ministère de l'Administration Territorial et de la Décentralisation (MATD-BF) 2005), is the county town of the department of the same name. The department is composed of 29 villages, among which the biggest in term of population is Sakora (1900 inhabitants) and the smallest, Nadrifa (142 inhabitants). The Karaboro form the most important ethnic group of the department of Tiefora. These are among the oldest groups that had settled in the south-west of Burkina Faso (Institut National de la Statistique et de la Démographie (INSD-BF) 1985). Agriculture occupies first place in the economic activities of the rural community of Tiefora in terms of source of income and occupation of the workforce. The practice of gardening is still underdeveloped. All the farmers practice extensive agriculture characterized by rainfed subsistence: low agricultural equipment, limited access to credit. They use less chemical fertilizer and organic matter (organic manure, crop residues, mulching).

3.2.3. Climate

The poor temporal rainfall distribution imposes the rhythm of agriculture. The site of Tiefora is under the influence of the Tropical Savannah climate with two well separated seasons (Société Générale d'Aménagements Hydro-Agricoles (SOGETHA) 1963, Peel et al. 2007). The average yearly rainfall that amounts to 1020 mm possesses a poor monthly distribution. The month of August is the most rainy (often more than 300 mm) and 96% of the rainfall occurs between April and October (Figure 3.2). Therefore, it is not surprising that most of the traditional crop production is done during the rainy season. Irrigation was introduced in the early 1960's. The monthly reference evapotranspiration ETo, obtained from the FAO Database Climwat for the station of Bobo-Dioulasso (Figure 3.1), plotted against the rainfall shows that full or supplemental irrigation is needed from November to May (Figure 3.3).

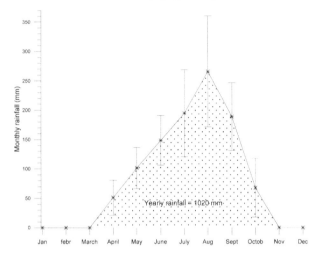

Figure 3.2: Average 1996-2009 monthly rainfall of Tiefora

Data source: Local Department of Agriculture. The error bars are standard deviations.

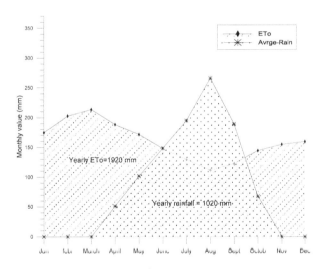

Figure 3.3: Rainfall versus ETo

The average ETo values are those of Bobo-Dioulasso, 100 km away from Tiefora but within the same Tropical Savannah (Figure 3.1). Sources: ETo : Climwat FAO; Rainfall: MAHRH

3.2.1. The dam

The characteristics of the dam are reported in Table 3.1 and Figure 3.4.

Table 3.1: Characteristics of the Tiefora dam

Structure type	Homogeneous earthen dam
Watercourse name	Lafondé
Construction enterprise	GRSFED-TP
Date of construction	1962
Restoration	1972
River basin (km²)	175
Maximum height (m)	4.5
*Dam top length/width (m)	780/3.0
Dam top topographic elevation (m)	271.00
Dam Upstream slope---Downstream slope	2/1---2.1
Dam Upstream covering/Downstream covering	Riprap masonry/Grass
**Spillway position/type	Central/Concrete
**Spillway length (m)/discharge (m³/s)	130/162
Spillway topographic elevation (m)	269.00
Gates (of left bank and right bank elevations) floor elevation (m)	266.20
Crest of main intake weir elevation (m)	267.40
Available water depth for irrigation (m)	269.00-267.40= 1.60 m
Estimated water available for irrigation	523 420 m³

*SOGETHA planned a dam of 920m, but 780m is the length measured in the field.
** Confirmed by field measurement in 2011.
Sources: (Société Générale d'Aménagements Hydro-Agricoles (SOGETHA) 1963,
Office National des Barrages et des Aménagements Hydroagricoles (ONBAH) 1987);
field measurements

Figure 3.4: Spillway of Tiefora earthen dam

Mr. Keïta discusses with Dr. Hayde about the influence of the spillway
overflows on the water logging of the irrigated valley located 920 m downstream.

3.2.5. The valley bottom irrigation scheme

The significant number of people living in Tiefora in the years 1960's led the project promoters to set four objectives. At the realisation of the irrigation scheme in 1963, the area of Tiefora was considered to have a zone with high population density, most of who were living below the poverty threshold. Hence, the scheme was planned to achieve the following objectives (Société Générale d'Aménagements Hydro-Agricoles (SOGETHA) 1963):

- increase the economic potential of the region;
- improve the local populations living conditions;
- familiarise the farmers with intensive agriculture and irrigation practice;
- regulate farmers' income not to be affected by annual the rainfall distribution.

Main canal

Water flows from the dam to the valley in a long earthen transport canal. Currently this, locally called *main canal*, has a length of 920 m. It starts with a stilling basin, located just downstream the dam. Before reaching the valley bottom rice field, this feeds three other offtakes irrigating a non-official crop vegetable growing area of 8 ha owned by villagers. The canal is not lined, it trees and shrubs on its sections and banks, forming a kind of gallery forest, thereby with higher roughness and suspended sediment concentration in the water. The cross section of the canal is trapezoidal. The average bottom width is 0.60 m while the top width is 1.00 m and the depth is 0.70 m. Hydrometric measurements made along the canal during the dry season (April 2010) in two control sections provided the following results: mean water depth: 20 cm, mean flow velocity: 0.80 m/s and an average discharge of 0.12 m^3/s. At the beginning of the valley bottom and just after the secondary canal offtake, the main canal water flows into an overflow canal.

Water distribution system components

The water distribution system inside the valley is composed of a set of canals and distribution structures. The system is equipped with a 640 m long secondary canal. This canal is lined with concrete. Its rectangular cross section has the following dimensions: bottom width 45 cm, average height 60 cm (Figure 3.6: Typical tertiary offtake on the secondary canal in Tiefora). Hydrometric measurements made along this canal in April 2010 in three different control sections provided the following results: average water depth 17 cm, average flow velocity: 1.05 m/s and an average discharge of 80 l/s. This canal feeds 21 tertiary canals. The intake structures at the head of these tertiary canals consist of metallic hand gates. The water level regulation is ensured by a small weir placed one meter downstream of the offtake. The tertiary canals are quasi-rectangular earthen canals. These tertiary canals feed the 39 farm plots having an average size of 0.11±0.03 ha. The dimensions of a tertiary canal are: 30 cm of width, 30 cm depth. Hydrometric measurements performed in April 2010 along 4 tertiary canals showed an average water depth of 12.5 cm, a velocity of 0.62 m/s and a discharge of 23.3 l/s.

Figure 3.5: Location of historical plains

One can see the Plain I downstream the dam, the Plain II of 25 ha commanded by the right bank gate and the Plain III 800 m downstream commanded by the left bank gate. The latest one was finally developed as the valley bottom irrigation field of Tiefora.

Figure 3.6: Typical tertiary offtake on the secondary canal in Tiefora

A transversal weir in the secondary canal rises the water level to its designed value and feeds the
secondary canal on the left with a discharge regulated by the metallic vertical gate

Drainage network and roads

The protection of the valley against flood and the circulation are ensured by two main
drains and a network of bunds. The drainage network consists of earthen canals
surrounding the valley bottom with a total length of about 1670 m. The drains are
trapezoidal with the following features: bottom width, 0.50 m; top width, 1.00 m; depth,
0.40 m. It should be noticed that there is no protection dike. These two drains collect the
water flowing from the secondary drains fed by the tertiary drains located in the farm
plots. Some 1670 m long small roads of three meters wide gives access to the valley.
Inside the valley, the tertiary canal protection bunds are used as paths to access the farm
plots. The crossing of the roads by the canals is arranged with small scuppers.

3.2.6. Topography and valley history

At its creation of the irrigation system in 1963, the designer (Société Générale
d'Aménagements Hydro-Agricoles (SOGETHA) 1963) described three areas in the
plain that could be developed taking into account the topography and the dispersions of
the watercourse meanders:

- the first area was called *Plain I* located immediately downstream the dam.
  'This area would not be equipped but is available for vegetable growing by
  water withdrawal from wells.' (Société Générale d'Aménagements Hydro-
  Agricoles (SOGETHA) 1963). The multitude of watercourse meanders and
  the smallness of available land were the main reasons for which this area was
  not developed;
- the second area was located on the right riverbank of the watercourse. This
  area with a potential of 25 ha was called Plain II and was also withdrawn.
  There was already a main intake gate of the right bank (Figure 3.5) for the

irrigation area but it was calculated that the reservoir will not have enough water for both this area and the third one located in a topographical better position downstream;

- the third and more convenient area named Plain III, was located 800 m downstream the dam, inside a valley bottom (Figure 3.5). That is the one that was developed into an irrigated rice project called Tiefora.

After careful topographic measurements, the Plain III was selected at that time to be the site of a modern irrigation system. This plain was described as a plain that was "constituted by a flat and regular bottom, [...] surrounded by the Lafondé watercourse and the foot of the hills" (Société Générale d'Aménagements Hydro-Agricoles (SOGETHA) 1963). A first estimate of the cross sectional average slope gives 0.0% in the valley bottom and confirms the SOGETHA's description (Figure 3.7). SOGETHA projected a neat area of 33 ha to be developed with a main canal discharge of 120 l/s. This discharge would feed four secondary canals named S1, S2, S3 and S4, each of which commanded a smaller area (between 8,00 and 10.10 ha). Field topographic measurements in June 2011 reveal a total of 39 official plots, with net plot sizes ranging from 0.22 ha to 0.54 ha. The total net area is about 16 ha.

3.2.7. The Pedology and the reasons of choosing Plain III

Even though already seen as a poor hydromorphic soil with iron threat in the years 60's, Plain III was selected for the development project for some good reasons. A first study performed by IRAT (a French agronomic research institute) at the beginning of 1960's showed that the soils of the valley bottom of Tiefora are *hydromorphic*, deep, homogeneous and non-gravelly. High content of clay revealed traces of waterlogging and oxido-reduction. (Société Générale d'Aménagements Hydro-Agricoles (SOGETHA) 1963).At that time, an important reason that advocated for the development project was the fact that in more than 60% of the valley bottom rice was grown in by local farmers without any modern irrigation system (Société Générale d'Aménagements Hydro-Agricoles (SOGETHA) 1963). Later in 1985, another pedological study was implemented by ONBAH/AGRAR-UND HYDROTECHNIK Gmbh in the area located between the valley bottom rice irrigation scheme and the watercourse Lafondé. The area investigated by ONBAH is not located in Plain II but they share the same boundary: the Lafondé watercourse. The study indicated that the area was homogeneous, 'part of the Lafondé right bank alluvial plain' with a unique type of soil (Office National des Barrages et des Aménagements Hydroagricoles (ONBAH) 1987). The soils are described by ONBAH as "*hydromorphic* with *pseudogley* (stagnosols) [soil horizon at shallow depth characterized by gray to beige tones, with many red spots or rust], developed on top of an alluvial material with fine texture to very fine texture". Hydromorphy (characterised by an oxygen deficiency) is moderated in surface (the soils are flooded several times during the rainy season – even if the flood period is generally no longer than 1 to 3 days – when waters from the spillway adds up to the discharge carried by the watercourse Lafondé). However, hydromorphy is accentuated in deeper layers, under the influence of the phreatic level raise.

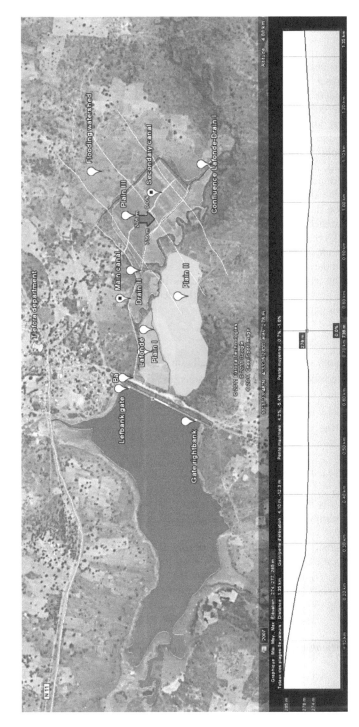

Figure 3.7: Middle cross section of the Tiefora valley bottom rice scheme

The slopes within the valley bottom are flat, generally nil. However, the slopes can be more than 5.0%, particularly on the eastern side watershed which is one of the 2 sources for the valley bottom flooding (the second is the overflow of the spillway, conveyed by Lafonde watercourse).

3.2.8. The problems

Flooding

Floods have been always part of the history of the irrigated valley bottom of Tiefora. The preceding observations of ONBAH about hydromorphy and flooding were in line with those of SOGETHA in 1963. Furthermore, more recent field pre-investigations performed by the current research project in 2010 confirmed that the plain is flooded two or three times during the rainy season, precisely in the month of August. The floods are described by the farmers as covering the downstream third of the irrigated area of Tiefora valley bottom and bringing fine sediments. Using farmers' testimony and the hydrography map, it appeared that the flood process is initiated from the point 'Confluence Lafondé-Drain I', at the junction of the two water courses (Figure 3.7). During August, the spillway overflow (Figure 3.4) is drained by Lafondé River and its waters meet those drained by Drain I (which is a protective drain collecting external runoff from the eastern river basin). These waters, flowing with difficulty in the rough Lafondé riverbed, create a backwater that fluxes from the 'Confluence Lafondé-Drain I' to the valley bottom. The consequences are that the third downstream area of the valley bottom is flooded during 2-3 days.

Though the current gross irrigated area is 16 ha versus the 33 ha projected by SOGETHA, the flooding problem was clearly identified in 1963. A simulation was done with the following two scenarios.

Scenario I

A decennial rainfall of *Pdec* = 125 mm falls on the valley bottom rice field and its surrounding watershed while, for some reasons, the check valve from the main drain (most likely Drain I on Figure 3.7) to the Lafondé *is closed*. It is assumed that the rice plots will be under water during not more than 36 hours. Thereafter the check valve will be opened and the drain has to convey from the valley the discharge calculated according to the following equations:

$$Vol = 10\, P_{dec} C_r S_{ws} \qquad\qquad (3.1)$$

where: *Vol* is the runoff volume of water collected at the drain outlet in m^3; P_{dec} is the decennial rainfall in mm; C_r is the runoff coefficient; S_{ws} is the watershed drained in ha.

This volume, coming from the watershed was spread over the gross area S_{field} = 40 ha of valley bottom projected field, yielding a water layer. This water layer was added to the Hl_{field} generated directly by the rainfall over the valley bottom. The total height had to be evacuated within a maximum of *Tmax* = 3 days by the main drain. The corresponding formula used was the following:

$$Q_{evac} = \frac{\dfrac{Vol}{S_{field}} + 10\, Hl_{field}}{86.4\, T_{max}} \qquad\qquad (3.2)$$

where: Q_{evac} (l/s/ha) is the average discharge to be evacuated by the drain within three days; *Vol* (m^3) is the volume flowing from the surrounding watershed to the valley bottom; S_{field} (ha) is the rough area of the valley bottom; Hl_{field} (mm) is the water layer generated by the decennial rainfall in the field; T_{max} (days) is the maximum time during which the field remains under water; and 86.4 is a unit conversion factor.

The numerical application was performed as it follows.

$$\left.\begin{array}{l} P_{dec} = 125; C_r = 0.60; S_{ws} = 150 \text{ ha} \rightarrow Vol \approx 110,000 \text{ m}^3 \\ S_{field} = 40 \text{ ha }; Hl_{field} = 125 \text{ mm}; T_{max} = 3 \text{ days} \end{array}\right\} \rightarrow Q_{evac} \approx 15 \text{ (l/s)/ha} \quad (3.3)$$

For an area of 40 ha, the corresponding discharge is 600l/s and the main drain was calibrated to convey this water to the Lafondé. One should notice that this scenario does not take into account the overflow volume from the spillway. It admits also a submersion duration of 3 days.

Scenario II

In the second scenario, a decennial rainfall of P_{dec} = 125 mm falling upon the valley and its surrounding watershed was also assumed. However, here the check valve from the main drain to the Lafondé *is open*. It was assumed that the drains directly convey water to the Lafondé. No submersion (or only a 'short submersion' of 6 hrs and 40 min) was allowed. The computations yielded a discharge of Q_{evac}= 5,000 l/s. This could lead to a drain "largely overdesigned in comparison with the size of the irrigation scheme". Finally, the first scenario was adopted: a flooding period of 3 days of the field is allowed (Société Générale d'Aménagements Hydro-Agricoles (SOGETHA) 1963).

Currently, flooding is a regular phenomenon considered as having a negative impact on rice production, even if these impacts are not well known. It especially happens during the month of August every year 2-3 times after heavy rainfall, when the spillway overflows and the drains waters meet downstream and create a backwater that runs to the valley (Figure 3.7). A consistent amount of fine materials were brought into the plain by the flood though their impacts on fertility and infiltration alteration were not yet investigated. The current research planned to address infiltration configuration within the valley bottom toposequence (see chapters 7 and 8).

A rice yield that can be improved

Though cultivar was developed by a local research institute, the historical record of the rice grown in Tiefora showed a trend to decrease. The rice grown is FKR19, a cultivar developed by IN.ERA[5] from the Nigerian strain Mashuri X IET 1444. This rice is basically produced from *Oryza Sativa*, the Asian rice. FKR19 was launched in Burkina Faso for the first time in 1984 (Institut de l'Environnement et de Recherches Agricoles (IN.ERA) 2000). The potential yield is 6-7 tons/ha. In Figure 3.8 the average rice yield evolution over 13 years in Tiefora, from 1997 to 2009 is presented. The values were

---

[5] Institut de l'Environnement et de Recherches Agricoles (French). Institute for Environment and Agricultural Research.

computed from the folder of the Chief ZAT[6], the technical agent on the Ministry of Agriculture in charge of the irrigated field. Furthermore, one can notice that the yield is generally slightly higher (0.50 ton/ha) in dry season than in the rainy season. A trend to drop also appears between 1997 and 2004. The later year was a drought period according the rainfall records (Kanté 2011), with a yield smaller than 3.0 tons/ha. However, as it can be noticed, the yields are low compared to the potential.

Several reasons could be invoked to explain this low yield in both dry and rainy seasons. The lengthy duration of the nursery period was a relevant issue invoked by the Chief ZAT, which statement was confirmed by some farmers (Figure 3.9). Field investigations allow eliminating number of these reasons. For example, it seems that fertilizer application is done according the recommended doses of 300 kg/ha of N-P-K (14-23-14) and 100-150 kg/ha of urea (Institut de l'Environnement et de Recherches Agricoles (IN.ERA) 2000, Sokona et al. 2010). Flood episodes of 2-3 days occur generally in August at least 3 times and may affect the downstream third of the valley bottom. Yet, the dry season, with no flood, has also a low yield. Therefore, floods seem not to be the main reason of such a yield level. Another probable cause to explain the low yield might be that the farmers do not comply with the cropping calendar. However, field investigation reveals a satisfactory cooperation between the farmers and the UAT[7] Chief, following up the cropping pattern (Sokona et al. 2010). Therefore, the low level of the rice yield was hypothetically associated with the soil quality and drainage issues to be investigated.

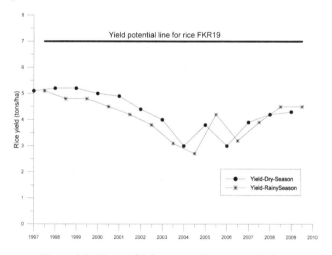

Figure 3.8: Rice yield drop over 13 years in Tiefora

Average yield over the 13 years and both dry and rainy seasons is 4 tons/ha

---

[6] Zone d'Appui Technique (French). Technical Assistance Zone. The agent is a worker of the Ministry of Agriculture. A ZAT coordinates the actions of several UAT.

[7] Unité d'Appui Technique (French). Technical Assistance Unit. The agent is a worker of the Ministry of Agriculture.

Figure 3.9: Nursery in Tiefora

Prof. Schultz (left) discussing with Dr. Dianou (microbiologist) about
the impact of nursery aging on rice development.

A soil with iron toxicity symptoms

Field investigations were performed in 3 cross sections in the valley bottom in May 2011 in order to bring about some soil properties. The cross sections were located upstream, middle and downstream respectively. During this pre-diagnosis, qualitative observations were made along the first 30 cm layer of the top soil and 12 samples were collected for size analysis. Though more investigations were needed to better characterize the area, the results (Figure 3.10) for this top layer indicate the dominance of clay. The average clay content is 44.0% vs. 32.3% of sand in the first 30 cm top layer. In May 2011, at the end of the dry season after the end of the irrigation period, the water table was located between 1.00 and 1.50 m according to the area. During the irrigation periods, independently from the season, (rainy or dry), the water table rises to almost ground level (Kanté 2011). Hydromorphy was visible in deeper layers with green (blue + yellow) colours, testifying the coexistence of reduction (Fe^{2+} = blue) and oxidation (Fe^{3+} = yellow). During higher water table periods, this phenomenon was expected to be more accentuated. The soluble Fe^{2+} was precisely the iron component the excessive absorption of which could lead to iron toxicity. Therefore, the current research project planned to address this issue of ferrous iron in the soil and its toxicity for the rice (See Sections 0, 0 and 0).

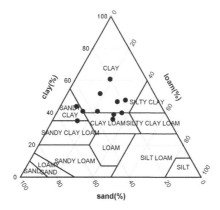

Figure 3.10: Size analysis in 3 cross sections of Tiefora valley bottom in May 2011

Triangle of USDA soil classification. Every dot represents a sample taken from the field either
upstream cross section or in the middle or downstream.

3.3.   The site of Moussodougou

3.3.1. Location and activities

Moussodougou is a small village located at the heart of the Tropical Savannah region of
Burkina Faso with a population living essentially from agriculture. The coordinates of
the village are 10° 49' 55" North and 4° 56' 15" West. The population was estimated to
some 2000 to 3000 inhabitants according to the farmers' estimate. This population
practices the production of mango which is mainly sold after transportation to the cities
of Bobo-Dioulasso (the economic capital) and Ouagadougou (the administrative capital
of Burkina Faso) located some 130 and 510 km away respectively. Though important
for occupation of manpower, mango fruits are usually paid at such a low price – one can
estimate the sell price at less than 10 FCFA/kg –that its impact on the revenues is not
enough to get out of the poverty trap. Therefore, most of the people living in the village
also practice agriculture and breeding.

3.3.2. The problems

In an ultimate effort to improve their revenue, a group of farmers initiated a call for help
to develop the valley bottom of Moussodougou into a rice irrigated system that quickly
ran into important difficulties. The call was addressed to a development programme of
Taiwan that helped the farmers to make de bunds and the farm plots and offered
fertilizers and rice seeds. Local cultivars, essentially FKR 14 and FKR 19, created by
IN.ERA (Institut de l'Environnement et de Recherches Agricoles (IN.ERA) 2000), were
proposed to farmers to grow. Spate irrigation is done using the floods of the river
flowing across the middle of the valley. The first crop growing season took place in
207-2008, over an area of about 30 ha composed of 0.21±0.1 ha farm plots. This first
production, according farmers testimony, did not give rise to any remarkable problem.
But, by the second production season of 2009-2010, rice started to show ferrous iron
intoxication symptoms, noticeable by the browning of the leaves. Very rapidly
afterwards, several farm plots production ended to zero yield due to iron intoxication.

The farmers organised themselves into a so called "pre-cooperative" (farmers' organisation) and tried to find various solutions together. They changed the rice cultivar without any improvement. They incorporated more organic fertilizer and that worsened the toxicity.

In 2012, the site of Moussodougou – one area among the 370 000 ha of valley bottoms in south Burkina Faso (Ministère de l'Agriculture de l'Hydraulique et des Ressources Halieutiques (MAHRH-BF) 1999) – was included in the current research project. It was considered as one of the sites facing severe iron toxicity where a substantial contribution could be made to reclaim the soils and improve rice production. This PhD research, though investigating only two farm sites of Tiefora (16 ha) and Moussodougou (35 ha), is addressing a wide threat to rice production – iron toxicity – threatening 60% of the traditional valley bottoms of Tropical Savannah in Africa according to Africa Rice (West Africa Rice Development Association (WARDA) 2006).

3.4.   Project components

In order to draw strong inferences for the research performed in Tiefora and Moussodougou and return to the farmers results that could be really beneficial to them, the methodology was built on the basis of the research hypotheses. This methodology consisted of two main components called C#1 for the first and C#2 for the second.

3.4.1. First component

The first component C#1 included in situ investigations focused factors having a potential influence on iron toxicity. Two valley bottoms irrigated rice schemes were selected: Tiefora and Moussodougou, both located at the heart of the Tropical Savannah region of Burkina Faso. The following aspects – directly related to the research questions and the hypotheses – were studied:
1. identification and mapping of all the iron intoxicated areas with the assistance of the farmers;
2. ground surface and in-depth soil sampling in order to study the distribution of the clay and ferrous iron in the valley bottom;
3. comparison of iron toxicity key factors – the clay distribution, the pH; the ferrous iron concentration – between a slightly and a heavily ferrous iron intoxicated valleys in order to highlight the most important risk factors exposing to iron toxicity;
4. characterisation of clay distributions within the valley bottom toposequence and proposition of appropriate subsurface drainage techniques able to alleviate iron toxicity;
5. investigation of the under-phreatic level permeability of the soils in connection with their location within the valley bottom toposequence;
6. comparison of the farmers' irrigation schedule and the assessment of those schedules obtained by water consumption records and time series analysis in order to propose improvements.

These in situ studies had at least two tremendous advantages. Firstly, they brought the researcher into close contact with the farmers and the Tropical Savannah rice irrigated valley bottoms, making subsequent visits more efficient and more focused. Secondly, the investigations provided new insight about the various factors – soil grain size, clay, acidity, depth, organic matter – which have a potential impact on iron toxicity. These key factors were afterwards used to design optimized controlled experiments in which the risks of influence of lurking factors – difficult to contain in real field because of the almost impossible replicability of treatment conditions within different farm plots – were kept at minimum. These experiments were implemented within the framework of the second component C#2 of this research.

3.4.2. Second component

The second component C#2 focused on subsurface drainage experiments in microplots grown rice in order to investigate some key factor impacts on iron toxicity. These microplots were made in concrete (Dianou 2005). PVC perforated pipes were installed at the bottom to collect infiltration through the soil after irrigation. The soils of the microplots were carried from the valley bottoms of Tiefora and Moussodougou. Two key factors – subsurface drainage and lime incorporation – for each of which having two treatments (or factor levels) were investigated. Measurements of the soil responses to the treatment conditions were performed directly in the microplots or on extracted soil samples. These measurements were: ferrous iron Fe^{2+} concentrations, soil acidity pH, oxido-reduction potential ORP, and dissolved oxygen DO at various depths are also performed. The development of iron toxicity symptoms were measured as IRRI iron toxicity scores (International Rice Research Institute (IRRI) 2002) during the entire crop growing season. The size of the microplots used by Jacq and Fortuner (1979) in their investigation on SRB activities was 1 m x 1m x 1m and they were under ground. Dianou (2005) used on ground concrete microplots of half height for his investigations. It appears that drainage is easier when the microplots are on ground as in the case of Dianou. Therefore, it was decided to modify the type of microplots used by Dianou by introducing in the system a surface drainage tap on one hand, and a subsurface tap associated with a perforated pipe on the other hand (Figure 3.11).

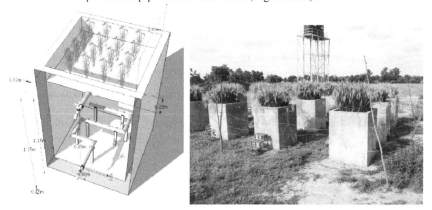

Figure 3.11: Concrete microplots with sub- and surface drainage system

The microplots are constructed in reinforced concrete to prevent cracks due to temperature changes following the alternation of the rainy and the dry seasons.

4. IRRIGATION SYSTEMS PREDIAGNOSES AND UPDATE

4.1. Diagnosing and mapping Tiefora valley bottom

At the beginning of the research activities, a mission over some 450 km led a research team from Ouagadougou to Banfora, in view of prediagnoses and information updating. The team visited the dam, and the irrigated rice valley bottom of Tiefora, located 25 km away from Banfora. The team realised that if all the 36 plots of this rice-growing area of 16 ha were not affected by the iron toxicity, it was clear that a certain number of farm plots were suffering of it. Farmers on the site were preparing their nurseries for the next rice season. The team noted in some nurseries the presence of ocher iron oxide significantly contrasting with the rest of the surface rather greyish clay (Figure 4.1). Presumably just below the ocher surfaces, iron would be found in the reduced state Fe^{2+}, already affecting the development of young rice. Based on these observations, it was recommended that all farm plots where the phenomenon of iron toxicity is noticeable will be carefully mapped. The rice yields would be followed up in these plots but also in sound plots in view of a comparison. Soil samples would also be extracted from the iron intoxicated plots of in situ measurement, but also to experiment in controlled devices. This preliminary diagnosis was implemented a few weeks later (Figure 4.2). A map censing all the iron affected farm plots in a detailed drawing produced served as basis for all the subsequent investigations in Tiefora (Figure 4.4).

Figure 4.1: Nursery areas plagued by iron toxicity in Tiefora

In areas where iron ocher is concentrated, one
can see that the young rice plants become scarce.

(A)   (B)

(C)   (D)

Figure 4.2: Prediagnosis and equipment installation at Tiefora and Moussodougou

(A) Extracted soil samples. (B) Installing E-water level diver. (C) Checking
the farm plots. (D) Inventorying the ferrous iron intoxicated farm plots

4.2.   Lessons from valley bottom irrigated rice fields of Valley du Kou

Following the works in Tiefora, another mission visited the ferrous iron intoxicated area
of the Kou valley in order to draw lessons from their fight against iron toxicity.
According to some preview research reports (Ouattara 1992), more than 300 ha were
severely affected by iron toxicity and abandoned by the farmers (Figure 4.3). The
research team noticed that the area was since used by IN.ERA (Institute of Environment
and Agricultural Research of Burkina Faso) for testing various rice cultivars as for their
resistance to iron toxicity. The mission was able to have discussions with the
management of the area. Based on these exchanges, on the visual observation of the
deficiency of the drains often clogged or obstructed by ponding water, it was suggested
that iron toxicity – also displaying reddish ocre colour clearly visible in the water and
soil – was here mainly due to inadequate drainage. This hypothesis was corroborated by
the fact that farms in adjacent plots had obviously holy and healthy rice, contrasting
sharply with the contaminated area. The so-called farmers' fields were well drained to
the pond. In the eyes of the mission, one of the most appropriate solutions to iron

toxicity would indeed be drainage, whether surface or subsurface. Farmers in adjacent plots had understood.

Figure 4.3: Red colour of the water showing the presence of ferric iron F^{3+} in this plot of Kou valley

Ferric iron Fe^{3+} in itself is not harmful for rice. However, when it remains for a long time in anaerobic conditions, it is reduced to Fe^{2+} which is soluble and can be absorbed by the roots of the plant. This leads to physiological disorders manifested by the appearance of brown and orange leaves. Rice yields may fall to zero, depending on the degree of severity of the toxicity.

4.3.   Diagnosing and mapping Moussodougou valley bottom

The work on the second research site of Moussodougou was implemented during a research team visit in three steps. Moussodougou was identified with the help of a department of the Ministry of Agriculture located in the region. It was rice-growing valley bottom of 36 ha mainly used in the rainy season when the lowland is flooded by a river passing through its mid and facing a severe iron toxicity. At the beginning, the team worked to confirm the draft map of the farming system initially designed by GIS from aerial photographs. With the support of the farmers, a march was conducted farm plot by farm plot to identify the true boundaries of the drawn plots and cancel inexistent bunds of separation from the map. The second step was the identification in the plots of the presence of symptoms of iron toxicity (red trihydroxide precipitated iron at the soil surface). It appeared that 90% of the areas identified were facing the problem of course with varying severities. The team eventually recorded, with the assistance of the farmers, the names of all the farmers of the organisation of Moussodougou and gave identification codes to their plots. Several maps were subsequently produced for the research activities (Figure 4.5).

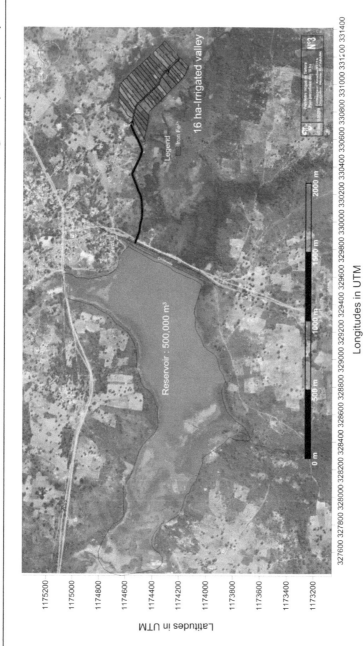

Longitudes in UTM

Figure 4.4: New farm plot map of Tiefora, with ferrous iron locations

A total amount of 15.2 ha of 39 farm plots) were delineated, 69% (11 ha) of which were confronted with a slight iron toxicity (Fe^{2+} smaller than 75 mg/l at depth 50 cm) and a pH greater than six. Ferrous iron location in the farm plots are indicated in red.

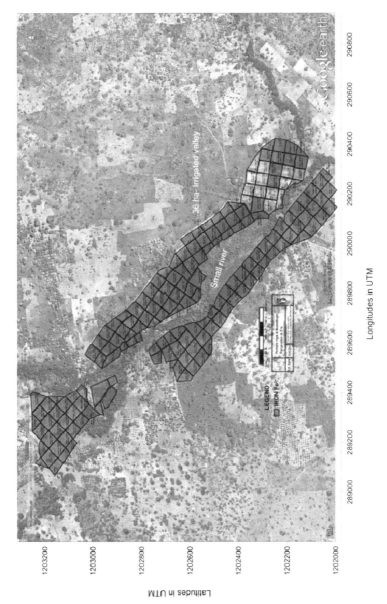

Figure 4.5: New farm plot map of Moussodougou, with ferrous iron locations

A total amount of 159 farm plots (35.2 ha) were delineated, 90% of which were confronted with a severe iron toxicity with Fe^{2+} often greater than 900 mg/l and a pH of 5-6. Ferrous iron location in the farm plots are indicated in red.

5. CLAY AND FERROUS IRON STRATIFICATIONS

In the tropical savannah valley bottoms, the classic salinity build up (excess of sodium) is not the main consequence of poor drainage. The main threat is ferrous iron toxicity (an excessive accumulation of iron in the plant, taking place especially when the soil is not well aerated) which also drastically reduces the crop yield, particularly rice. The West Africa Rice Development Association (WARDA, now called "Africa Rice") estimated that at least 60% of the swampy cultivated inland areas of Africa are affected by varying degrees of iron toxicity (Sahravat et al. 1996). The same association reported that in West Africa, iron toxicity causes up to 12% through to 100% of rice yield drops, depending on its severity and the tolerance of the rice variety.

The predominance of gleysoils (Schaetzl and Sharon 2005, Ogban and Babalola 2009) in the valley bottom in tropical Savannah reduces crop yield in many irrigation schemes. This is particularly true when the soils are exposed to several cycles of flooding during the year with high groundwater table like in Tiefora, a small 16 ha-irrigated scheme located at the heart of the tropical Savannah region of Burkina Faso. Tiefora valley bottom rice has attracted various populations looking for livelihood. However, the rice yield remains as low as 3.0 tons/ha (Sokona et al., 2010). This low yield thought to be a direct consequence of ferrous iron toxicity (Kanté 2011).

In order to provide means for alleviating iron toxicity, various researchers brought contributions to the characterization of valley bottom redoximorphic soils (Kessler and Oosterbaan 1974, Barron and Torrent 1986, Ogban and Babalola 2009). However, previous research does not provide detailed information about the soil stratification and ferrous iron distribution within the soil (Jackson and Sherman 1953). This potential stratification can help building a strategy against iron toxicity. In the current research, we have hypothesized that both clay content and ferrous iron Fe^{2+} concentrations are stratified in the soil, at least under certain conditions. To check these assumptions, a randomized block sampling followed by a statistical analysis was implemented in the case of the typical tropical Savannah valley bottom of Tiefora.

5.1. Soil sampling and measurements

The sampling was performed in view of finding the relationships among four variables: the depth in the soil, the percentage of clay in the soil, the ferrous iron Fe^{2+} concentrations and the *pH*.

The V-shape of the Tiefora valley bottom suggested making a randomized block sampling (Boslaugh and Watters 2008). Within a cross section, three conditions of slope exist: left bank (LB), valley axis (VA) and right bank (RB) (Figure 5.1). These slopes are suspected to impact on the clay deposition in the valley. In addition, the longitudinal slope is also suspected to impact on clay deposition in the soil profile. Hence, the valley was longitudinally divided into an upstream region (UR), a middle region (MR) and a downstream region (DR) (picture). The blocks were thus defined within the cross section, each related to a homogeneous slope region. These blocks are LB, VA and RB. In order to integrate the longitudinal effect, one replicate was done in each of the 3 regions UR, MR and DR. Therefore, the valley was divided into nine sub-regions and a soil extraction site was selected randomly from each one. This way the confounding effects on clay deposition of the longitudinal and lateral slopes were assumed neutralized, since each sloping condition is equally represented in the final sample of soil extracts.

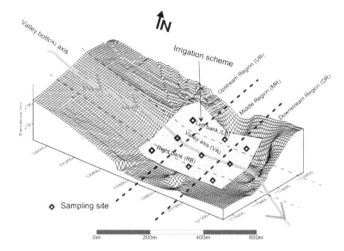

Figure 5.1: Valley bottom randomized block sampling sites

One can see the different sloping conditions from upstream to downstream and from the left to the right bank. This let to dividing the valley bottom into nine sampling sites. For each site, several soils extracts were taken at various depths. The numbers on XY axes are the UTM coordinates

Two series of soil extraction were conducted in the flooded soil. In the first series, taking into consideration the fact that rice root maximum depth is 100 cm (Allen et al. 1998), three depths were targeted: 30 cm, 50 cm and 100 cm. Therefore, nine boreholes were perforated using a boring auger and a soil extract taken from each of the three depths. This led to 27 soil extracts. The second series of operations bored also nine holes in the randomly selected sites. However, the depths were not limited to 100 cm since the purpose was to investigate on soil stratification. While boring, an extract was taken when a change in colour or texture was observed. Finally the borehole execution was ended each time a fine sand layer – forming a stratum of water inflow– was found. The second series led to the gathering of some 51 soil extracts.

These two sets of soil extracts (27+51) were used for two different purposes. The 27 soil extracts were used to investigate clay percentage and ferrous iron concentration in the soil. As for the 51 soil extracts, they were utilized to look further for clay percentage stratification in depth, up to 5 meters when nothing was revealed in the 100 cm of top soil. With the 27 extracts, the ferrous iron concentrations were measured using a reflectometer with Fe^{2+} reflectoquant strips (Persson 1997). The strips used measure concentrations within the two ranges [0.5, 20mg/l] and [20, 200 mg/l] in the soil solution. Dilution factors were made any time the observed concentrations in ferrous ions were out of theses ranges. The soil solution was obtained through dilutions with factor 5 from the saturated extracts (Murray 1994). For both groups of soil extracts, the clay proportions were obtained by grain size and sedimentometry analyses according to ASTM (American Society for Testing and Materials).

5.2.   Statistical analyses

It is anticipated that stratifications according to the depth will occur with the two response variables, clay percentage and ferrous iron concentration. Therefore ANOVA test (Keïta et al. 2013a) was performed with the software Minitab to look for significant differences in means within the populations underlying the various soil extracts. The three conditions of applicability of ANOVA – the normality of the underlying populations, the equity of variances and the equity of the number of samples – were checked before implementation (Boslaugh and Watters 2008). These conditions were met with the 51 extracts of soils. However, the important difference in variances did not permit the application of ANOVA to the 27 other soil extracts. Consequently, the Welch t-Test that does not require the equality of two sample variances (Welch 1947) was applied in this case to look for stratifications in ferrous iron concentrations.

5.3.   Clay stratification occurrence

One-way ANOVA test can be applied with the 51 soils extracts despite the differences in the number of observations (Table 5.1) according to the depths. This test is known to be more sensitive to non-normality of the populations and non-equity of the variances than to the differences in number of observations (Table 5.1)(Persson 1997). In normal conditions, the materials in a savannah valley bottom are mainly composed of colluvial and alluvial depositions in the form of strata (Foth 1990). However, in the case of Tiefora, the land has been irrigated more than 50 years and one would expect that agricultural modelling of the soil for rice growing would destroy any stratification, at least areas close to the top soil. Hence our null hypothesis is that there is no stratification in the soil, and the alternative hypothesis is that there exists stratification in terms of clay percentage. ANOVA was applied to test these hypotheses at 5% of significance. In addition, a Fisher test about strata differentiation was also applied.

Table 5.1: Clay percentage of the soil extracts taken in the valley bottom

Depths (cm)	N observations	Mean %Clay	StDev	Variances	Min	Max
00-100	25[(*)]	28.9	12.3	150.7	3.6	56.4
100-200	9	19.6	11.4	129.2	1.1	36.6
200-300	5	18.18	10.4	108.3	3.1	31.0
300-400	4	15.8	11.6	133.8	4.1	26.9
400-500	6	13.8	07.0	48.5	6.0	23.1

[(*)]Without 2 outliers

Table 5.2: ANOVA - the mean clay% are significantly different for at least two depths

Source	DF (degree of freedom)	SS (sum of squares)	MS (mean squares)	F of Fisher	P-value
Depths class (cm)	4	1805	451	3.47	0.015[(*)]
Error	44	5726	130		
Total	48	7531			

[(*)]Significant at $\alpha = 0.05$

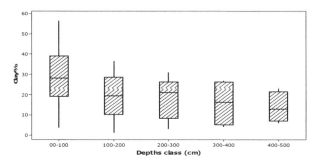

Figure 5.2: Clay percentage dispersion and decrease with the soil depth

Note: The box-and-whisker plots – created by John Wilder Tukey – constitute a powerful tool for
multiple comparisons of population means (Tukey 1977, Hoaglin et al. 1991). These plots
represent five quantities within the data set. First, $\%Clay_{min}$ and $\%Clay_{max}$ are at the bottom and
the top of the whisker line. Secondly, the lower line of the rectangular box represent the first
quartile $Q1$ (25% of the sample values are smaller than the value of $Q1$). Thirdly, the line inside
the box represents the median $Q2$ (50% of the sample have a lower value than $Q2$). Finally, the
top limit of the rectangular box represent the third quartile $Q3$ (25% of the sample elements have
a value greater than $Q3$. Thus, the box contains 50% of the data of the sample. The box-and-
whisker plots provide two very useful checks in statistics: the normality check and the
significance of (apparent) differences in sample means. In effect, if the median line ($Q2$) in the
rectangular box is *a symmetry axis*, then the sample is more likely drawn from a population with
normal distribution. Furthermore, for two small samples having *at least 5 data* each one and
coming from normal populations, if the *boxes do not overlap* when compared side by side, then
the two populations means are significantly different (*p-value* < 5%). For two samples having *at
least 30 data* each one and coming from normal populations, if the *median line of one sample
does not intersect the box of the other one* when extended, then the corresponding two
populations means are significantly different (*p-value* < 5%). (Mathews 2005).

   The results show that clay content in the soil reduces steadily but definitely with
the depth (Figure 5.2). Because of the small p-value (1.5%), we reject the null
hypothesis that all depths are equivalent in terms of clay content and that there is no
stratification (Table 5.2). The picture given by the soil according to the strata ID (3rd
and 4th column in Table 5.3) is that the clay percentage is significantly higher in the
first 100 cm than at any other depth below. However in the layer 200-300 cm, though
the mean clay percentage is lower than in the top 100 cm, it shares some similarities
with the latest (sharing of soil stratum letter A, see Figure 5.2 and Table 5.3).

   Table 5.3: Fisher method - the mean *Clay%* are different for at least two depths

Depths class (cm)	Number of observations	Mean of clay percentage	Soil stratum ID[**]
00-100	25[*]	28.9	A
100-200	9	19.6	B
200-300	5	18.1	AB
300-400	4	15.8	B
400-500	6	13.8	B

[*] Without 2 outliers

[**] Strata that do not share a common soil stratum ID letter are significantly different

Clay content drops with the depth in the soil may explain why metallic ions (such as ferrous ion Fe^{2+}) and other chemicals tend to concentrate more near the soil surface (Herzsprung et al. 1998). In fact, since clay and humus (also found near the surface of soil) are one of the most chemically active components of the agricultural soils, most of the cations and anions will stick near in the top soil (Foth 1990). However, if the soil is quite permeable, e.g. not composed of heavy clay, seasonal fluctuating ground water table may bring the ions deeper in the soil during dry periods (Peel et al. 2007). There would be one exception: if the valley is also irrigated during the dry period. In that case, the exchangeable ions may remain in the top soils. This hypothesis was checked for ferrous iron in the case of Tiefora as examined below.

5.4.   Ferrous iron concentration in the rootzone

One of the most reliable ways to test the hypothesis of ferrous iron stratification in the 100 cm top soil was to apply the two samples Welch t-Test (Welch 1947), which does not require any assumption about equality of variances. The one-way ANOVA could not be performed because of the great difference within the variances (Table 5.4 and Figure 5.3). Both the Bartlett's and the Levene's tests of equality of variances (Carroll and Schneider 1985) performed at 5% of significance. This led to the rejection of the equalities of the variances. Therefore Welch t-Test was performed at 5% of significance level. The null hypothesis is then that the three soil depths were equal in terms of mean ferrous iron concentrations. The test led to the rejection of this null hypothesis in favour of the stratification with a p-value equal to 0.0% (Table 5.4 and Table 5.5).

In fact, it seems that either high or low ferrous iron concentration can be associated with either low or even neutral *pH* level in the soil (Table 5.6). High ferrous iron content in flooded rice soils has often been found in association with an low *pH* of less than 5, after the oxidation of pyrite (Moormann and Breemen 1978, Suryadi 1996). In Tiefora, all the three layers have a slight but equal acidity of 6.5, and one could expect less ferrous ion Fe^{2+} in all of them. That means that the high ferrous iron concentration (994 mg/l; Table 5.4) in the depth of 30 cm was rather unexpected since associated with only a slightly acid *pH* of 6.5. At the same time, a comparatively very low ferrous iron concentration is found at 50-100 cm depths for the same *pH* value. Therefore, it seems that high Fe^{2+} concentration can occur in either acidic or neutral conditions even though the more acid the medium, the more soluble – and then potentially toxic for the rice – is Fe^{2+} (Suryadi 1996).

Subsurface drainage of valley bottom in Tropical Savannah

Table 5.4: Ferrous iron concentration in the soil extracts taken at three depths

Depth (cm)	N observ	N* missing	Mean Fe^{2+} (mg/l)	StDev	Variance	Minimum	Median	Maximum
30	8	1	994.1	215.3	46363.3	615	1025	1200
50	9	0	74.56	10.93	119.53	55	80	85
100	9	0	72.33	18.11	328	43	73	98

Table 5.5: Welch two samples t-Test for Fe^{2+} at three depths

Couple of depths	Difference of means (mg/l)	95% Confidence interval	t-Test of difference	Deg. of freedom	p-Value	Depth (cm)	Iron stratum[**]
30 vs. 50 cm	919.6	[739;1100]	12.07	7	0.00[*]	30	A
30 vs. 100 cm	892.4	[682;1103]	10.04	7	0.00[*]	50	BC
50 vs. 100 cm	2.22	[-13;17]	0.32	13	0.76	100	CB

[*]Significant at $\alpha = 0.05$

[**]Strata that do not share a common letter are significantly different

56

Table 5.6:*pH* variation in the soil at 3 depths

Depth (cm)	N observ.	N* missing	Mean pH	StDev
30	9	0	6.4	0.4
50	9	0	6.3	0.6
100	9	0	6.6	0.6

The iron distribution profile of the top 100 cm soil presents some similarities with other cation distributions. The high iron concentration in the 30 cm top soil surprisingly contrasts with the horizons underneath (Figure 5.3). From an average of 994.1 mg/l it drops to 47.5 mg/l at the depth 50 cm, and then slightly rises to 72.3 mg/l (Table 5.4), all this occurring in the clay dominant section of the soil. Such a drastic change is difficult to explain by the intake of iron by the rice roots that remain after harvest in the soil. It seems reasonable to think that this concentration in the top 30 cm is due, not to oxidation which would have conducted to Fe^{3+} compounds (Dent 1986), but to evaporation following capillary rise in the clay layer. Therefore, these phenomena would have similar effect to the one occurring in soils affected by sodium salinity (Bajwa et al. 1986) in which white sodium chlorite (NaCl) is visible during dry period on the surface of the soil. In fact, in Tiefora, a thin layer of ferric iron Fe^{3+} (the oxidized form) is also visible with its vivid red colour in rice nursery areas when no ponding water remains.

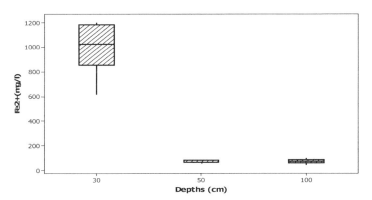

Figure 5.3: Ferrous iron concentration dispersion with the soil depth

Ferrous iron alleviation techniques will have to take into account this stratification in the valley. It is noteworthy that within this non-stratified clayey 30 cm-top soil, even though the variation is important (variance = 46360 mg²/l²), the ferrous iron concentration is very high everywhere in the flooded and oxygen depleted valley bottom (mean ~ 994 mg/l). There is no single location where the phenomenon is not present. This is also in favour of the explanation that water evaporated leaving behind the ferrous iron. Subsurface drainage in permanently wet conditions (leaching) with open channels (Schultz 1988), or buried pipes implemented in the 100 cm to layer may help to oxygenate the root zone and transform soluble ferrous iron into solid ferric iron, less toxic for rice.

5.5. Conclusions

The stratification of clay and ferrous iron concentration can help to develop a strategy to alleviate iron toxicity in tropical savannah soils under irrigated rice. In our case study, the results show that clay content is significantly higher in the 100 cm top soil, with a mean of 28.9% (Stdev of 12.3%). It drops to 13.8% at 400 cm of depth. We have found that ferrous iron was mainly concentrated within the top 30 cm of the clay, reaching 994 mg/l (Stdev of 215). The iron concentration drops much more quickly than the clay since already at 50 cm depth, the value is only 74.5 mg/l (Stdev of 10.9). Moreover, this high iron concentration is observed within the top 30 cm of soil all over the flooded valley bottom, supporting the idea that it is was left behind after capillary rise and evaporation. These extreme stratification and concentration in the top soil provide a way to alleviate iron toxicity. It is proposed to maintain permanent wet conditions during the growing period in the irrigated lands in combination with leaching (subsurface drainage) in the fallow periods. This may help removing the excess of ferrous iron or preventing its occurrence.

6. HIGHER IRON TOXICITY RISK IN SINGLE-SEASON IRRIGATINON

Rice is a global cereal facing yield growth challenges in several regions. It is currently consumed by more than 3 billion people and is cultivated on more than 163 million ha (Nguyen and Ferrero 2006, (FAO) 2013). Food security challenge is closely linked to rice production since this cereal is essentially self-consumed, being internationally marketed for only 7% (Wailes 2005). Nevertheless, rice yield is either stagnant or declining in several regions since 1990s (United Nations Food and Agriculture Organization (FAO) 1996). In fact, if the current production trend had to continue, the cereal food availability estimated to 150 kg/person in 2005 will drop to 130 kg/person by 2035 (Schultz et al. 2009). About 10% of the crop farming areas are found in valley bottoms in the Tropical Savannah of West Africa, where exploitation focuses on rice production but faces also a yield decline (Masiyandima et al. 2003). The West African rice production only covers 50% of the needs and thus most of the countries in the region are rice importers (Lançon and Benz 2007). Nigeria, Senegal and Ivory Coast are among the biggest importers (Wailes 2005). In Burkina Faso, the importation was estimated to more than 210,000 tons in 2010 (Institut National de la Statistique et de la Démographie (INSD-BF) 2011, Dembele et al. 2012). Iron toxicity is one of the most important constraints to yield growth in Tropical Savannah valley bottoms, threatening up to 60% of the swampy area, with a yield reduction of 10-100% (West Africa Rice Development Association (WARDA) 2006). In Burkina Faso, 300 ha ferrous iron intoxicated soils were abandoned in the Vallée du Kou in 1986, most of which remained uncultivated up to date (Ouattara 1992).

Iron toxicity issue in West African Tropical Savannah is currently addressed by various fields of research. In agronomy, Africa Rice devotes an important effort in the development of ferrous iron resistant cultivars (West Africa Rice Development Association (WARDA) 2006, Somado et al. 2008). Microbiology strives to identify and quantify the bacterial activities involved in the redox processes, occurring in the rice irrigated valley bottoms, in order to propose chemical and/or biotechnological means of reclamation (Ouattara 1992, Dianou 2005). Still, one cannot find in the current research any study comparing a traditional single-season spate irrigated rice soil with a double-season modern irrigation system in terms of acidity and ferrous iron concentrations threats. Such a comparison can lead to a better understanding of the processes and to different alleviation strategies against iron toxicity. Our study was carried out taking this view.

6.1. Soil sampling

The procedure strove to avoid bias while collecting soil extracts from these colluvio-alluvial valley bottoms. Both valleys possess longitudinal and transversal slopes making possible the occurrence of different thicknesses of clay or concentration of ferrous iron for example. Therefore a randomized block sampling was used (Mason et al. 2003, Rumsey 2009). The valley was divided into three zones longitudinally and three zones transversally. The longitudinal axis was assumed to be the line of the main irrigation canal passing through the middle of the valley in Tiefora, while it coincides with the river bed in Moussodougou. Considering the symmetrical shape of this second valley, the soils extracted from the rightbank of the river are assumed to present similar properties than the corresponding investigated area on the leftbank (Figure 6.1). In Tiefora, the soil extracts were taken from both sides of the asymmetric valley (Keïta et al. 2013a). Twenty seven soil extracts were collected from each of the two valley bottoms.

6.2. Measurements

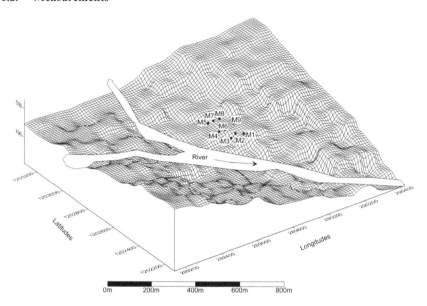

Figure 6.1: Moussodougou valley with the "M" soil sampling point locations

A randomized block sampling was implemented. Three soil extracts were collected from 30, 50 and 100 cm depth in each random "M" point inside a "block". The blocks represent different longitudinal and transversal sloping conditions. The corresponding area by symmetry on the right bank of the river was assumed to have similar sloping conditions.

With a strong concern about data accuracy, one set of measurements was implemented in the field and a second set in the laboratory. For the field side, the *pH* and ferrous iron concentrations were measured shortly after extraction from the flooded rice soils and their introduction into anaerobic boxes. For this purpose the extracts were diluted 5 times to get the soil solutions (Murray 1994). A *pH* meter was then used to determine the current acidity in the samples, and a reflectometer to quantify Fe^{2+} concentrations (Keïta 2013b). For the laboratory side, grain size sorting using standardized American Society for Testing and Materials (ASTM) sieves and sedimentometry yielded the clay proportions *(%Clay)* for each depth according to US Department of Agriculture (USDA) classification. In addition, the organic matter content *(%OM)* was assessed by weighting and oven drying at 900 °C. Finally, the dry bulk density *(Db)* at each individual depth was obtained by weighting and oven drying at 105 °C.

6.3.   Statistical analysis

Since one of the objectives was to compare Tiefora and Moussodougou on a sound ground, statistical hypotheses testing was required (Montgomery and Runger 2011). Prior to the choice and the implementation of the tests, three conditions had to be verified: i) the equality of the number of samples, ii) the normality of samples underlying populations, and iii) the equality of variances. These checks were performed with the previous five response variables (Fe^{2+}, pH, $\%Clay$, Db and $\%OM$) for both sites and at each of the depths. When the three conditions were fulfilled, a One-Way ANOVA+Fisher grouping could be applied to check if there was any significant difference at the depth under consideration for the two sites. The depth is considered as the independent variable. The null hypothesis was that the five response variables are equivalent when the two sites are compared depth per depth. In case one of the prerequisite conditions was missing, a non-parametric hypothesis testing model was applied (Boslaugh and Watters 2008).

6.4.   Geochemistry

A geochemical analysis endeavoured to find the underlying causes of the differences revealed by the statistical hypothesis testing. This analysis considered the chemical reactions related to iron oxido-reduction as well as the bacterial involvement as catalysts (Fredrickson and Gorby 1996, Emerson et al. 1999). Tropical Savannah soils are predominantly hematite (Fe_2O_3) and precipitates of iron(III)-hydroxide ($Fe(OH)_3$). Therefore oxido-reduction processes are alternatively occurring in the wet valley bottom soils (Fisher and Schwertmann 1974, Moormann and Breemen 1978, Spaargaren and Deckers 2004). The key reactions, particularly in hematite dominant environment, were selected and examined in order to predict their impact on the increase and/or decrease of the two main response variables in iron toxicity, i.e. Fe^{2+} concentrations and the acidity indicator pH (Conklin Jr. AR 2005). Consequently, a table was drawn, showing each important reaction involving hematite or iron(III)-hydroxide with its impact on the change of ferrous iron concentration and the acidity. Corresponding to the reactions in this table, crop growing season – wet or dry – in Tiefora and Moussodougou were inserted according to the assumed oxido-reduction process occurring. This analysis led to uncovering the reasons behind the differences between the two sites and to practical proposals to prevent or reduce iron toxicity and acidity.

6.5.   Conditions check for hypotheses testing

The first two conditions required for enabling the application of One-Way ANOVA – equality of sample numbers and normality – were rather well met using Anderson-Darling normality statistic (Boslaugh and Watters 2008). All the samples of the five response variables have eight or nine elements. If the significance level is set $\alpha = 5.0\%$, all the samples are normally distributed except at 3 depths for 2 variables. These depths are 50 cm/pH and 100 cm/$\%OM$ in Tiefora, and 100 cm/$\%OM$ in Moussodougou. The related p-values of Anderson-Darling statistics are slightly smaller than 5% (Figure 6.2). However, this does not constitute a real obstacle to the use of One-Way ANOVA since this test is less sensitive to departure from normality (Montgomery and Runger 2011).

The equality of variances check constituted a greater obstacle to the application of One-Way ANOVA. The computations performed for the 5 response variables at the three depths in Tiefora and Moussodougou showed that the samples of Fe^{2+} and of %Clay provided Levene p-values smaller than 5%(Gastwirth et al. 2009, Rumsey

2009) Therefore, the variances are not equal in both Tiefora and Moussodougou for these two variables at the related depths. Consequently, a hypothesis test model less sensitive to the equality of variances was applied, namely the non-parametric Welch 2-sample t-Test (Welch 1947, Brase and Brase 2007).

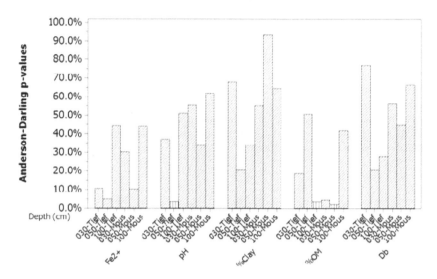

Figure 6.2: Normality check for the 5 response variables

The p-value of the Anderson-Darling statistics are significant at α = 5% except for the three depths with vertical fill pattern (50 cm/*pH* and 100 cm/%OM for Tiefora; 100 cm/%OM for Moussodougou). Except these three depth-variables, other samples had a normal distribution.

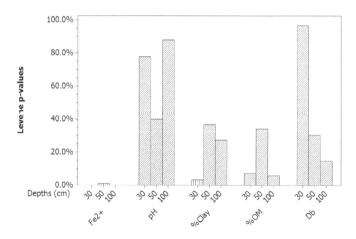

Figure 6.3: Equality of variances check for the five response variables

The Levene statistic p-values are greater than 5% for three response variables – *pH*, *%OM* and *Db* – but are smaller for Fe^{2+} at all depths and *%Clay* at depth 30 cm (vertical fill pattern of the bars). For these four depths, the related variables are not equivalent in variance in the two sites of Tiefora and Moussodougou (Levene 1960).

6.6. Statistical results

The application of hypotheses testing for comparing the two sites – Welch 2-samples t-Test, One Way ANOVA+Fisher Grouping – do not lead to significant differences except for Fe^{2+} and the pH. Both sites were similar in the corresponding depths when considering the clay content (%Clay), the organic matter content (%OM) and the dry bulk density (Db). For example, the organic matter content for all the three layers are equivalent in both sites, with an average varying between 7 and 8% and a p-value of 87% (Table 6.1 and Table 6.2). Similarly an average 1.3-1.5 dry bulk density was found equivalent for both site, and the clay content %Clay was found not significantly different with the averages varying from 20-40%.

Table 6.1: One-way ANOVA: OM (%) versus Depth (cm) in Tiefora and Moussodougou

Source	DF	SS	MS	F	p-value
Depth (cm)	5	19.5	3.9	0.37	87.0%[*]
Error	48	506.5	10.6		
Total	53	526.1			

S = 3.2; R² = 3.70%; R²adj = 0.00%

[*] p-value greater than 5% ; all the %OM values are similar in Tiefora and Moussodougou

Table 6.2: Grouping Information Using Fisher Method

Depth (cm)	N	Mean(%OM)	Grouping[*]
030-Tief	9	8.0	A
030-Mous	9	7.6	A
050-Tief	9	7.4	A
050-Mous	9	6.1	A
100-Tief	9	7.0	A
100-Mous	9	7.2	A

[*]Means that do not share a letter are significantly different. The organic matter content is equivalent in both sites Tiefora and Moussodougou (same letter A).

Regarding the iron concentrations, the results show that Moussodougou is more exposed to ferrous iron intoxication than Tiefora. The application of Welch 2-samples t-Test to compare the ferrous iron Fe^{2+} concentrations at the three individual depths in the two sites brings about two important differences (Figure 6.4). First, while Fe^{2+} concentrations were not significantly different in the 30 cm top soil in both sites, they were much higher in Moussodougou than in Tiefora at depths 50 and 100 cm. The values were some 1100 mg/l higher in Moussodougou (significant at $\alpha = 5\%$), and even more than 1800 mg/l higher at depth 100 cm (Table 6.3). The second remarkable aspect was that while the ferrous iron gets a quite uniform concentration (no significant p-value) in the whole 100 cm soil profile in Moussodougou, it suddenly drops to less than 100 mg/l underneath 30 cm in Tiefora (Keïta et al. 2013a). Such discrepancies are revelatory of different biochemical reactions that will be addressed in the geochemical analysis section below. At least the high concentrations show why the iron toxicity is a bigger issue – reaching 50% of the farm plots – in Moussodougou than in Tiefora (10% of the farm plots).

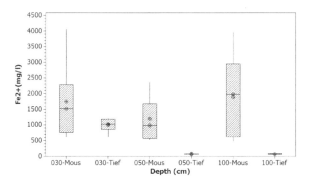

Figure 6.4: Comparison of ferrous iron concentrations on the sites of Tiefora and Moussodougou

While the Fe^{2+} concentrations were quite uniform in Moussodougou, they dropped in Tiefora below 30 cm depth. Only the top 30 cm had comparable ferrous iron content in both sites.

Table 6.3: Tiefora versus. Moussodougou -- Welch 2-samples t-Test for Fe^{2+} at 3 depths

Depths	μ_{Mousso}- μ_{Tiefo}(mg/l)	95% Confidence interval	t-Test of difference	Deg. of freedom	p-Value	Iron stratum[**]
30 cm	750	[-100;1600]	2.0	8	7.7%	A-A
50 cm	1110	[601;1625]	5.0	7	0.1%[*]	B-A
				8	0.2%[*]	
100 cm	1820	[855;2780]	4.3			C-A

[*] Significant at $\alpha = 5.0\%$. The corresponding Fe^{2+} values are significantly different in Tiefora compared to Moussodougou. [**] The first letter stands for Tiefora and the letter for Moussodougou. Fe^{2+} concentrations are comparables at 30 cm on both sites. But at 50 cm and 100 cm, Fe^{2+} concentrations are much higher in Moussodougou than Tiefora.

The second response variable that made a real difference between the two sites was the acidity indicator, the *pH*. As the result of the application of One-Way ANOVA+Fisher grouping, it comes again that the top 30 cm draws the attention: Moussodougou with its average 5.7 is significantly more acidic than Tiefora (p-value = 0.4%). In the layers underneath till 100 cm, it remains more acidic (Table 6.4 and Table 6.5). However, compared to for example pyrite soils (*pH* 2-4) exposed to oxidation, the *pH* values were rather moderately acid, although the value can decrease to almost 4 in some layers, particularly in Moussodougou (Figure 6.5) (Attanandana and Vacharotayan 1986, Jacq and Ottow 1991). Though this is not surprising due to the hematite dominant environment, it leads in combination with a high ferrous iron concentration to a more severe iron toxicity in Moussodougou (score 7) (International Rice Research Institute (IRRI) 2002).

Table 6.4: One-way ANOVA: *pH* versus Depth level (cm)

Source:	DF (degre of freedom)	SS (sum of squares)	MS (mean squares)	F of Fisher	P-value
				4.08	0.4% [*]
Depth (cm)	5	7.79	1.557		
Error	48	18.3	0.381		
Total	53	26.1			

S = 0.617; R² = 29.90%; R²adj = 22.50%
(*) Significant at α = 0.05.

Table 6.5: Fisher grouping applied to *pH* in the two sites of Tiefora and Moussodougou

Depth (cm)	Nbr of observ.	Mean	Grouping
030-Tief	9	6.4	A
030-Mous	9	5.7	C
050-Tief	9	6.3	AB
050-Mous	9	5.7	C
100-Tief	9	6.6	A
100-Mous	9	5.8	BC

Means that do not share a letter are significantly different.

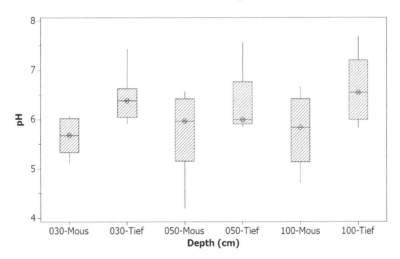

Figure 6.5: Comparison of pH on the sites of Tiefora and Moussodougou

6.7.   Geochemical analysis

The statistical analysis of the variables making a significant difference between Tiefora (a double-season irrigation scheme) and Moussodougou (a single-season spate irrigation scheme) brings about two important questions: i) Why is Moussodougou more "ferrous" and more acidic (and thus more exposed to iron toxicity) than Tiefora? ii) How ferrous was iron mainly concentrated in the 30 cm topsoil of Tiefora? To address these questions, some biogeochemical analysis would be helpful.

Though sulphate might be present to some extent, oxido-reduction processes related to iron were the most important in hematite dominant soils, such as found in the valley bottom soils of Tropical Savannah (Bouwer and Jackson 1974, Ouattara 1992). Acid sulphate soils – more often found in coastal floodplains and mangroves (Pons et al. 1982, Attanandana and Vacharotayan 1986) – are known to be very acidic when exposed to oxidation. The *pH* can decrease to less than three, when exposed to drainage for example. In Tropical Savannah inland valley bottoms, like Tiefora and Moussodougou, the main reactions involve the oxidation and the reduction of hematite (Fe_2O_3) and the precipitation of Iron (III) hydroxide $Fe(OH)_3$. These reactions may be catalyzed (up to million times) by iron reducing or oxidizing bacteria. The predominance of various iron (III) species may also be *pH* dependent (Ponnamperuma 1972, Emerson et al. 1999, Wenjing et al. 2008). The principal reactions involved valleys oxido-reduction are:

- Oxidation of pyrite: Simultaneous increase of Fe^{2+} and acidity (Suryadi 1996):

$$FeS_2 + 14Fe^{3+} + 8H_2O \rightarrow 15Fe^{2+} + 2SO_4^{2-} + 16H^+ \qquad (6.1)$$

- Precipitation of Fe(III) hydroxide: Simultaneous decrease of Fe^{2+} acidity (Rickard and Luther 2007):

$$4Fe^{2+} + O_2 + H^+ + H_2O \rightarrow Fe(OH)_3 + 3Fe^{3+} \qquad (6.2)$$

- Formation of hematite (oxidation): Decrease of Fe^{2+} and increase of acidity (Schwertmann U. and Murad E. 1983, Kato et al. 2008):

$$4Fe^{2+} + O_2 + 4H_2O \rightarrow 2Fe_2O_3 + 8H^+ \qquad (6.3)$$

- Reduction of Hematite: Increase of Fe^{2+} and decrease of acidity (Breemen 1992):

$$2Fe_2O_3 + CH_2O + 8H^+ \rightarrow 4Fe^{2+} + CO_2 + 5H_2O \qquad (6.4)$$

The impacts of the processes – reported in equations 1 to 4 – on the changes in ferrous iron concentrations and the acidity measured by the *pH* are of special interest. It is noteworthy that the velocity of these changes can be boosted million times by iron oxidizing (IOB) and iron reducing bacteria (IRB) (Fredrickson and Gorby 1996, Emerson et al. 1999). The oxidation of pyrite (Eq.(6.1)) when exposed to air or leaching leads to a strong simultaneous increase of both Fe^{2+} and acidity. Pyrite is mainly formed in marshy tidal mangroves and coastal floodplains with very low *pH* (Moormann and Breemen 1978, Suryadi 1996). In fact, a *pH* below 3.5 suffices to attest the presence of sulphuric acid resulting from pyrite oxidation (Attanandana and Vacharotayan 1986). For this reason, it seems reasonable to admit that pyrite is not much in the soils of Tiefora and Moussodougou having a moderate acid *pH* when exposed to air and which soils were weathered from hematite (Table 6.6). The precipitation of iron III hydroxide (Eq.(6.2)) leads to the simultaneous decrease of acidity and ferrous iron. This phenomenon was observed particularly in Moussodougou (Rickard and Luther 2007). It is essentially observable during spate irrigation in Moussodougou, but also during the two-season irrigation periods in Tiefora (Table 6.6).

Table 6.6: Dominant iron redox reactions in valley bottoms

Refer.	Reactions	Fe^{2+}	Acidity	Site
Eq.(6.1)	Oxidation Pyrite + IRB [(*)]	↗↗↗	↗↗↗	
Eq.(6.2)	Precipitation of Fe(III) hydroxide + IOB [(**)]	↘	↘	Mousso wet Tiefo wet season
Eq.(6.3)	Oxidation of Hematite + IOB [(**)]	↘	↗	Mousso dry season
Eq.(6.4)	Reduction of Hematite + IRB [(*)]	↗	↘	Mousso wet season Tiefo wet season

[(*)] IRB = Iron reducing bacteria   [(**)] IOB = Iron oxidizing bacteria
↗ = increase;   ↘ = decrease

From Eq.(6.3) and Eq.(6.4), it can be drawn how the redox processes result in a higher ferrous iron concentration and a more acidic environment in Moussodougou than in Tiefora. The oxidation of hematite will particularly occur in the soil of Moussodougou during the dry season (height months). During that period, ferrous iron concentration decreases, but the acidity increases (Eq.(6.3) and Table 6.6). This observation is confirmed by oxidising subsurface drainage experiments on rice growing microplots with hematite soils (Otoidobiga 2012). However, during the wet season with spate irrigation the reduction processes in Moussodougou resume on the hematite (Eq.(6.4)), leading to an increase of the ferrous iron concentration. The degree reached by this increase suggests an important activity iron reducing bacteria (IRB) (Wenjing et al. 2008). This high level of bacterial activity may be linked to river use as water resource for irrigation. The increase of Fe^{2+} concentration in a medium already more acidic due the previous dry season makes the soluble ferrous iron more dominant in the soil of Moussodougou compared with Tiefora. Finally, we suggest that the high Fe^{2+} concentration in the 30 cm top soil of Tiefora (the only one similar to Moussodougou) is due to a higher IRB activity in this layer.

Water management aiming to alleviate iron toxicity in Tropical savannah valley bottoms will have to deal with these three processes: i) precipitation of Fe(OH)$_3$, ii) oxidation hematite and iii) reduction of hematite. Fortunately, the precipitation of Fe(OH)$_3$, which is one of the most spectacular effects observable at the resurging points in the valley, is actually an ally in this fight. It is the only one simultaneously decreasing both Fe^{2+} and the acidity. It is often observable in open drains in the valleys. Therefore, open channel subsurface drainage will contribute to improve the fertility of the soils. In the same line, oxidation of hematite by bringing oxygenated water in the rice rootzone will decrease ferrous iron concentration, unfortunately along with an increase of acidity in short run. An incorporation of limestone (CaCO$_3$) will certainly reduce acidity in short run.

In very severe cases of iron toxicity, biotechnology resources can be investigated. A few studies have already been executed to identify the type of iron or sulphate reducing bacteria in Tropical Savannah valley bottom irrigated rice fields (Butlin et al. 1949, Ouattara 1992, Dianou 2005). There is a need for more investigations in this climatic zone about the positive impact of iron oxidizing bacteria (Thiobacillus ferrooxidans, Ferrobacillus ferrooxidans ...) in terms of reinstatement of ferric iron Fe^{3+} and the reduction of acidity.

Finally, it appears that practicing two-season rice growing presents less risk to develop iron toxicity in the valley bottoms. Therefore, the numerous traditional valleys where farmers are practicing a single-season spate irrigation would have to be developed into at least a double-season modernised rice irrigation scheme.

6.8.   Conclusions

There is a higher risk of iron toxicity development in a single-season compared to a double-season valley bottom irrigated rice in Tropical Savannah hematite rich soils. On the contrary of pyrite oxidation that leads to a simultaneous increase in Fe^{2+} concentration and acidity in costal floodplains and mangroves, the oxidation of hematite decreases Fe^{2+} but moderately increase soil acidity in Tropical Savannah. However, very high ferrous iron concentration may also occur (more than 3000 mg/l). We found a ferrous iron concentration 750-1800 mg/l higher in the single-season valley bottom in the 100 cm top soil profile. The pH, with an average of 5.7, was also significantly more acidic in the single season valley bottom, making ferrous iron more available in solution. During the dry season in which no crop was grown and the soil dried, acidity increased in the hematite. Therefore, our results show that there is a well founded interest to modernize traditional single-season spate irrigated rice into modern double-season schemes, equipped with an efficient open/pipe subsurface drainage system.

7. CLAY DISTRIBUTION AND ADAPTED DRAINAGE

Valley bottom soils are among the first ones used spontaneously by farmers in West-Africa to grow rice, though growing adverse conditions occur year after year (Nguu et al. 1988, Dembele et al. 2012, Keïta et al. 2013b). Several millions of hectares of rice growing valley bottoms face stresses and toxicities due to anoxic conditions (Becker and Asch 2005). For example, the West Africa Rice Development Association (WARDA, now called "Africa Rice") estimated that at least 60% of the swampy cultivated inland areas of Africa are affected by varying degrees of iron toxicity (Sahravat et al. 1996). The average rice yield has been estimated to 2.5-3.0 tons/ha for irrigated rice and this figure is even lower in Burkina Faso (Ouattara 1992). Various researchers brought important contributions to the characterisation of valley bottom waterlogged soils by using different parameters (Kessler and Oosterbaan 1974, Barron and Torrent 1986, Ogban and Babalola 2009). For example, Barron and Torrent (1986) attempted to provide visual observation criteria for waterlogging. Other parameters of valley bottom soils linked to infiltration processes such as the texture and the saturated hydraulic conductivity *Ksat* were studied (Kessler and Oosterbaan 1974), often more in view of classification of valley bottoms and their crop land suitability mapping.

Even though clay plays a key role (International Rice Research Institute (IRRI) 1985) in aerated water circulation through valley bottom soils, previous research in Tropical Savannah of West Africa remained silent about its distribution. Hardly can be found any study providing detailed information about clay spreading within the soil and its connection with subsurface drainage issues. The current research considers such connections by applying non-linear regression analysis to describe clay distribution, and proposes the most appropriate drainage technique for each situation met (Keïta et al. 2014a).

7.1. Soil sampling

In the 50 years old valley bottom, anticipating a heterogeneous soil requiring several samplings, two main types of boring machines were prepared. When passing through a clayey and cohesive but non-saturated soil, the Edelman auger with 70 mm of diameter was used. Across saturated soils under the groundwater table, in layers with sand or gravelly non-cohesive material, the Riverside auger of 70 mm diameter was used. Extension bars of 1 meter length with coupling sleeves were connected to the auger at the bottom and were operated on the top using a screwable handle. Nine boreholes of 2-6 m depth were made, three located in each of the upstream, middle and downstream cross sections (Figure 7.1). A total of 51 samples were collected using an undisturbed soil sampler tool any time there was a change in visual colour or in the structure (scratching the fingers) of the soil, all along the borehole. The number of samples varied from five to eight, according to the previous criteria. The collected soil samples were taken to the laboratory for grain size analyses.

The US Department of Army's classification (United States Department of Agriculture (USDA) 1997) was used to isolate clays particles smaller than 0.002 mm (2 μm). The processing according to this classification obeys the following rules: clays are particles smaller than 0.002 mm (2 μm), silt are grains with diameter between 0.002 and

0.02 mm (20 μm) and all grains between 0.02 mm and 2mm are total sand (fine sand + coarse sand). Between 2 mm and 11 mm are gravels, and above 11 mm are classified stones.

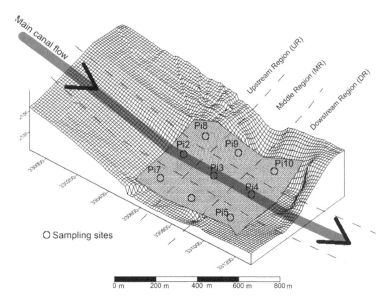

Figure 7.1 : The three cross sections and the location of the nine boreholes of Tiefora valley bottom

The circles represent the boreholes realized in the three areas delineated by the three cross sections. Pix represents the name of the borehole. The cross sections are located some 300 m from each other (Keïta et al. 2013a).

7.2.   Statistical analysis

With the aim to determine the most suitable regression curve for the clay distribution in the soil, the statistical analysis compared two non-linear models: the cubic and the quadratic. This analysis was performed using the statistical software package *GraphPad Prism v5*. The program makes a statistical comparison between the regression curves on the basis of the coefficients used in their equations (Keïta 2014).

Three criteria were mainly used to compare the two models. The goodness of fit of the model was the first. When the curves were equivalent in this regard, the simplicity of the model was the next criterion. The last criterion consisted of the practical value of the parameters in terms of subsurface drainage. The complexity of sample extractions led to the limitation of their number per borehole to less than ten. Despite of this limitation, the goodness of fit of the models was assessed based on three parameters. The first was the curve goodness of fit evaluated using the coefficient of determination R^2 and the standard deviation of the residuals S_{yx} (Montgomery and Runger 2011). Secondly, the test of normality of the residuals according to Shapiro - Wilk variables was applied (Shapiro and Wilk 1965). Finally, the Runs test method was used to evaluate how much the points deviate from the best fit (McWilliams 1990).

Once the most suitable regression curve was determined, the analyses strove to compare the clay profile similarities in the soil. Taking into account the sedimentation in toposequence inside the valley, which in the current case is from downstream to upstream, ((Keïta et al. 2014c)), three transversal cross sections were defined according the locations of the nine boreholes. Consequently, an attempt was made to find a common regression model for each of the three cross sections. This comparison was based on the hypothesis testing of the regression equation parameters. When the hypothesis testing revealed significant differences for the entire set of parameters, the location of the maximum of the regression curves in the soil - which influences the type subsurface drainage technology to be adopted - was checked to conclude about similarities of the soils (Härdle and Mammen 1993, Neumeyer and Dette 2003).

7.3. Soil texture

The colluvial and alluvial materials in the valley bottom reveal the predominance of clay in the top soil. For example, in the upstream cross section profile shown in Table 7.1. clay combinations dominate the first meters of the soil. Based solely on the grain size tables, it appears that the clay minerals percentage generally decreases as the depth increases. Clay content is particularly higher within the first meter in the soil. In this layer, the percentage of clay minerals is generally exceeds 20% and then decreases to less than 10% around 5 m. The phenomenon is more visible in the regression analysis performed below. One can also notice that while the clay content decreases, the sand content increases (Table 7.1, column 7). In fact, the fine sand proportion reaches more than 40% around 5 m in most of the boreholes. Though one cannot predict what type of soil will be found after the sand layer, it appears that the impervious layer is located rather in the top than in the bottom, as shown in the associated USDA soil class reported in the last columns of Table 7.1, Table 7.2 and Table 7.3. The first meter of the top soil is mainly formed by very fine material of clay loam and silty clay; these two classes having a common border in the USDA soil texture triangle. Remarkably, these results are contrary to the configuration – pervious layer overtopping impervious layer for example in peat soils – generally found in literature (Christen and Ayars 2001).

7.4. Finding a regression model

The attempt to find a unique equation for all the nine boreholes was indeed statistically successful. The equation found was quadratic and expressed as follows:

$$Clay(\%) = 30.6 - 7.3d + 0.9d^2 \qquad\qquad (7.1)$$

where: d represents the depth expressed in meters. In the previous quadratic equation the coefficients are the following: $B_0 = 30.6$; $B_1 = -7.3$ and $B_2 = 0.9$. The 90% confidence intervals for each of the three coefficients of the quadratic equation were respectively 25.9 to 35.3 for B_0, -12.90 to -1.70 for B_1 and -0.20 to 1.90 for B_2 for all the nine boreholes. Based on a null hypothesis "One curve for all data sets" and an alternative hypothesis of "Different curve for each data set", the hypothesis testing comparing all the similar coefficients for each of the nine boreholes yielded a p-value 45%, greater than the significant level of 5%. Therefore, the above unique regression curve Eq. (7.1) could be "accepted" (Boslaugh and Watters 2008) .

Table 7.1: Upstream cross-section boreholes grain size distribution

Sample No	Depth start (m)	Depth end (m)	%Clay	%Silt	%Gravel	%Sand total	%Total mater.	USDA Soil class (Triangle)
Piezo :	Pi7A	Geo coordinates UTM 30P :			X	330592	Y	1174470
39	0	0.1	25.8	25	9	49.2	100	Sandy clay loam
40	0.1	0.3	28.1	38.4	0.1	33.5	100	Clay loam
41	0.3	0.8	39.1	31.6	4	29.3	100	Clay loam
42	0.8	2	10.7	36.1	0.2	53.2	100	Sandy loam
43	2	4.3	7.4	37.4	0.2	55.2	100	Silt
44	4.3	4.7	21	41.3	0.2	37.7	100	Loam
45	4.7	5.3	17.9	47.7	0	34.4	100	Loam
Piezo :	Pi2A	Geo coordinates UTM 30P :			X	330734	Y	1174547
1	0	0.1	23.3	39.2	10	37.5	100	Loam
2	0.1	0.3	30	43	1.1	27	100	Clay loam
3	0.3	0.5	17.1	50.1	0.1	32.8	100	Silt loam
4	0.5	1.3	25.4	32.8	0.1	41.8	100	Loam
5	1.3	4	7.8	34.1	0.9	58.1	100	Sandy loam
Piezo :	Pi8A	Geo coordinates UTM 30P :			X	330864	Y	1174659
51	0	0.1	18	35	0.1	47	100	Loam
52	0.1	0.3	16.9	51.7	0.2	31.4	100	Silt loam
53	0.3	0.8	12.7	49.3	1	38	100	Loam
54	0.8	2.2	21.7	38.3	0.2	40	100	Loam
55	2.2	3.3	26.9	33.5	0.4	39.6	100	Clay loam
56	3.3	4.4	13.1	23.3	0.3	63.6	100	Sandy loam

The table contains data and computed results after grain size analysis. In the first column are reported the samples' ID as they were taken from the soil, the number representing actually the number carved on the sampling inox box. The second and the third columns data present the starting and the ending of a layer with identical texture and color. From column 4 to 7 are shown the grain distribution size calculation results from sieve analysis. In the column 9 – which is the last in the table – are shown the USDA soil triangle analysis results that determine the class of soil the layer belongs to. In the three internal lines of the table are shown the name of the piezometer and its geographical coordinates. The accuracy of the GPS measurements were 3 m. This description of the table is similar for all the nine boreholes.

Table 7.2: Middle cross-section boreholes grain size distribution

Sample No	Depth start (m)	Depth end (m)	%Clay	%Silt	%Gravel	%Sand total	%Total mater.	USDA Soil class (Triangle)
Piezo :	Pi6Abis	Geo coordinates UTM 30P :			X	330877	Y	1174344
33	0	0.1	20.3	38.4	0	41.3	100	Loam
34	0.1	0.3	38.9	29.1	0.4	32	100	Clay loam
35	0.3	1.4	36.6	31.1	0	32.3	100	Clay loam
36	1.4	2.4	13.6	37.6	0	48.8	100	Loam
37	2.4	4.7	23.1	39.8	2.5	37.1	100	Loam
38	4.7	6	20.3	30.4	1.3	49.3	100	Loam
Piezo :	Pi3A	Geo coordinates UTM 30P :			X	Y	330949	1174419
6	0	0.1	31.5	42.4	0.3	26.1	100	Clay loam
7	0.1	0.3	31.4	47	0.7	21.6	100	Clay loam
8	0.3	1.3	-	45.3	0	38.3	100	Loam
9	1.3	2.2	21	24.9	0.3	54.1	100	Sandy clay loam
10	2.2	2.7	3.1	35.4	0	61.5	100	Sandy loam
Piezo :	Pi9A	Geo coordinates UTM 30P :			X	Y	3331113	1174514
46	0	0.1	28	41	0.1	31	100	Clay loam
47	0.1	0.3	37	44.2	0.1	18.8	100	Silty clay loam
48	0.3	0.8	56.4	23.4	0.1	20.2	100	Clay
49	0.8	2.9	31	42.4	1	26.6	100	Clay loam
50	2.9	5.1	15.4	18.4	0.1	66.2	100	Sandy loam

Table 7.3: Downstream cross-section boreholes grain size distribution

Sample No	Depth start (m)	Depth end (m)	%Clay	%Silt	%Gravel	%Sand total	%Total mater.	USDA Soil class (Triangle)
Piezo :	Pi5A	Geo coordinates UTM 30P :			X	331114	Y	1174143
25	0	0.1	37.1	35.7	0.1	27.2	100	Clay loam
26	0.1	0.3	42.3	43.2	0.1	14.5	100	Silty clay
27	0.3	0.7	3.6	24.8	0.1	71.6	100	Sandy loam
28	0.7	1.6	32	32	0.2	36	100	Clay loam
29	1.6	3.2	4.1	24.3	0.2	71.6	100	Sandy loam
30	3.2	3.8	24.6	22.7	0.1	52.7	100	Sandy clay loam
31	3.8	4.4	12.5	21.9	0	65.6	100	Sandy loam
32	4.4	5	6	28.7	0	65.3	100	Sandy loam
Piezo :	Pi4A	Geo coordinates UTM 30P :			X	331181	Y	1174212
11	0	0.1	10.6	33.4	0	56	100	Sandy loam
12	0.1	0.3	40.6	47.2	0.2	12.2	100	Silty clay
13	0.3	0.5	42.1	45.3	0	12.6	100	Silty clay
14	0.5	1.4	25	26.8	6.1	48.2	100	Sandy clay loam
15	1.4	1.9	1.1	22	8.6	76.9	100	Loamy sand
Piezo :	Pi10A	Geo coordinates UTM 30P :			X	331289	Y	1174320
46	0	0.1	26.6	30.7	2.3	42.7	100	Clay loam
47	0.1	0.3	22.1	36	1	41.9	100	Clay loam
48	0.3	0.7	43.4	16	4	40.6	100	Loam
49	0.7	1.2	19.4	18.3	26	62.3	100	Silt loam
50	1.2	1.5	9.8	14.2	40.3	76	100	Silt loam

Due to important dispersion of the data points and the necessity of finding a physical meaning to the equations, this unique non-linear regression curve was not finally kept. The coefficients of determination ranged from -3.0 (at borehole Pi8) to 0.6 (at borehole Pi3), with three out of nine negative values (Rumsey 2009). The absolute sum of squares varied from 200 to 6000, indicating a quite important dispersion of data points around the regression curve (Boslaugh and Watters 2008, Montgomery and Runger 2011). Finally, the inherent heterogeneity of the clay distribution within the profiles and the toposequential deposition process of the sediments suggested a crosssectionnal approach (Figure 7.1), from upstream to downstream (Selley 2000). The regression curve coefficients are reported in Table 7.4.

7.5.   Regression goodness of fit

Despite of limited number of data points - 5 to 8 - per borehole (Table 7.5) the goodness of fit based on the three sets of parameters mentioned above can be considered as good. The coefficient of determination R^2 is greater than 0.5 (Brase and Brase 2007) for all the boreholes, except Pi5 and Pi6. However, particularly in a medium with high heterogeneity potentials, it is important to notice that goodness of fit cannot be limited to R^2 (Boslaugh and Watters 2008). Therefore, taking the analysis further, it appears that the normality test of the residuals led to Shapiro-Wilk coefficient W p-values ranging from 27 to 38%, all values greater than the significance p-value of 5%. This confirms the goodness of the fitting of the regression curves. Furthermore, the Runs test display also a balanced distribution of the data points around the regression curves (McWilliams 1990, Härdle and Mammen 1993), with "not significant" deviations (Table 7.5).

7.6.   Soil hydraulics analysis

The shape of the curves and the proportions of clay within the first two meters of the soil present several similarities, but also differences in the valley (Figure 7.2). It appears that despite of the various equations, clay proportions remains relatively high - between 20 and 30 % - within the 2 m topsoil in both upstream and middle areas. This is in line with the permeability values, found smaller in these two cross-sections - 0.10 ± 0.10 cm/h- compared with downstream (Keita and al.,2014. "Infiltration Rate Increase from Upstream to Downstream in a Valley Bottom", *International Agrophysics*, in press). In the last area where most of the pedological homogeneity is found, and particularly in piezometers Pi4 and Pi10, a clay proportion peak of 30-40% is reached at around 1m, while it remains smaller before this depth.

The clay thickness and location are two of the most important factors that help in the choice of the type of subsurface drainage for the soil. When the peak followed a more pervious layer such as in the downstream area (Pi4 and Pi10, Figure 7.2), but neither too close to the soil surface nor deeper than 2 m, the classical Hooghoudt subsurface design can be applied (Coles 1968, Ritzema et al. 2008). Perforated pipes in the case of Tiefora can be introduced around one meter depth within the soil (Christen and Ayars 2001). However, the persistence of a thick clay layer in the topsoil of the upstream and middle cross sections calls for application of alternative techniques such as "mole drainage" or permeability improvement practices (Schultz 1988). Mole drainage technique has been successfully applied against waterlogging and salt adversities for wide varieties of clay (Spoor et al. 1982, Cannell et al. 1984, Robinson et al. 1987). Elsewhere, crack making trenches were realised in the heavy, but non swelling clay (mainly illite) of the 165,000 ha agricultural land of IJsselmeerpolders in the Netherlands. This converted the almost impervious soil into a much lighter fertile medium, with an hydraulic conductivity of more than 100 mm/day (Schultz 1988). In Tiefora the range of 10 to 40 meq/100g of CEC displays the soils as mainly composed of illite (Sokona et al. 2010) in which these techniques may be applied.

Table 7.4: Regression parameters with 90% confidence interval

Piezo	Pi8	Pi9	Pi10	Pi2	Pi3	Pi4	Pi7	Pi6	Pi5
Selected regr	cubic	quadratic	quadratic	quadratic	quadratic	quadratic	quadratic	quadratic	quadratic

Coefficients with 90% confidence interval

B_0	20.2 ± 0.6	36.2 ± 31.0	19.2 ± 32.3	23.6 ± 16.8	32.8 ± 5.9	12.2 ± 34.8	33.6 ± 15.4	31.3 ± 21.3	33.3 ± 22.2
B_1	-17.7 ± 7.2	7.7 ± 40.6	47.5 ± 107.1	2.0 ± 31.9	-6.7 ± 13.1	76.4 ± 108.7	-12.2 ± 19.1	-4.0 ± 20.9	-8.4 ± 26.5
B_2	13.0 ± 3.9	-2.9 ± 3.0	-36.0 ± 64.1	-1.5 ± 7.4	-1.6 ± 4.6	-45.0 ± 54.1	1.8 ± 3.5	0.4 ± 3.39	0.8 ± 5.3
B_3	-2.1 ± 0.6	-	-	-	-	-	-	-	-

B_1, B_2...B_3 represent the coefficients of the quadratic (or cubic) regression equation $Clay(\%) = B_0 + B_1 d + B_2 d^2 + B_3 d^3$

Table 7.5: Goodness of the models applied to clay proportions

Piezo	Pi8	Pi9	Pi10	Pi2	Pi3	Pi4	Pi7	Pi6	Pi5
Selected regression type	cubic	quadratic	quadratic	quadratic	quadratic	quadratic	quadratic	quadratic	quadratic
Goodness of fit									
Degree fo Freedom	2	2	2	2	1	2	4	3	5
Regression coefficient R^2	1.0	0.6	0.7	0.7	1.0	0.8	0.5	0.2	0.3
Absolute Sum of Squares (Ab SS)	1.5	383.7	185.3	89.2	1.0	243.1	351.9	403.5	1129.0
Standard deviation of residuals ($S_{y,x}$)	0.86	13.85	9.62	6.68	1.00	11.03	9.38	11.60	15.03
Normality test of residuals									
Omnibus K2 of D'Agostino & Peasrson p-value	-	-	-	-	-	-	-	-	-
W of Shapiro-Wilk p-value	38%	41%	64%	74%	83%	41%	14%	31%	27%
Runs test									
Points above the curve	3	2	3	2	2	3	2	3	5
Points below the curve	3	3	2	3	2	2	5	3	3
p-value (runs test)	100%	90%	100%	100%	100%	90%	100%	90%	93%
Deviation from the model	Not significant	Not significant	Not significant	Not significant	Not significant	Not significant	Not significant	Not significant	Not significant

Null hypothesis = the most suitable regression is a quadratic function (second order polynomial);
Alternative hypothesis = the best fit function is a cubic function (third order polynomial).
The locations of Pix boreholes are given in Figure 7.1.

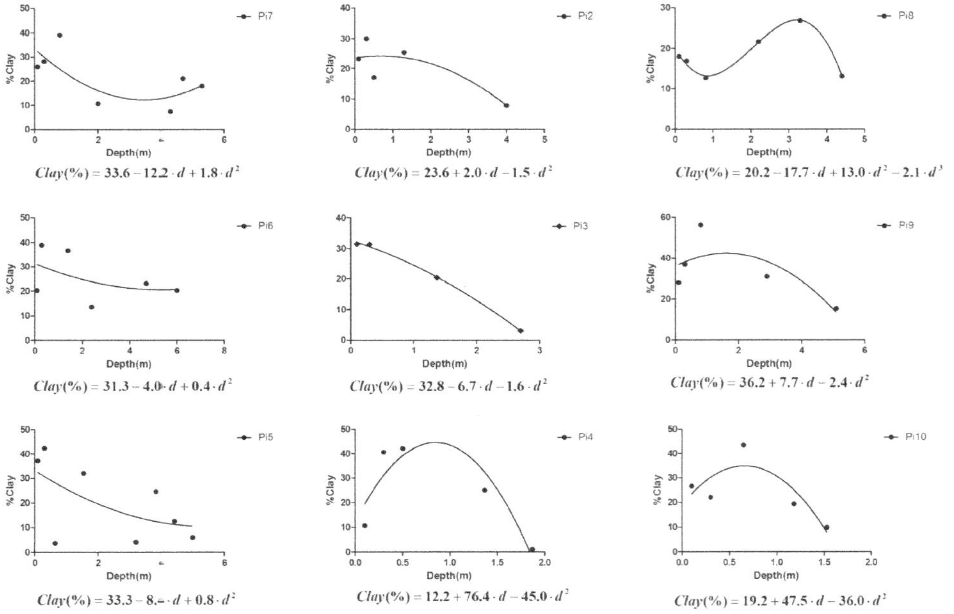

Subsurface drainage of valley bottom in Tropical Savannah

$Clay(\%) = 33.6 - 12.2 \cdot d + 1.8 \cdot d^2$

$Clay(\%) = 23.6 + 2.0 \cdot d - 1.5 \cdot d^2$

$Clay(\%) = 20.2 - 17.7 \cdot d + 13.0 \cdot d^2 - 2.1 \cdot d^3$

$Clay(\%) = 31.3 - 4.0 \cdot d + 0.4 \cdot d^2$

$Clay(\%) = 32.8 - 6.7 \cdot d - 1.6 \cdot d^2$

$Clay(\%) = 36.2 + 7.7 \cdot d - 2.4 \cdot d^2$

$Clay(\%) = 33.3 - 8.4 \cdot d + 0.8 \cdot d^2$

$Clay(\%) = 12.2 + 76.4 \cdot d - 45.0 \cdot d^2$

$Clay(\%) = 19.2 + 47.5 \cdot d - 36.0 \cdot d^2$

Figure 7.2: Regression best-fit curves and equations of the boreholes

The same names *Pix* were used for the boreholes (Figure 7.1) and the above regression curves. The three top curves Pi7 Pi2 and Pi8 are in the upstream area; Pi6, Pi3 and Pi9 in the middle, and Pi5, Pi4 and Pi10 are in the downstream area. The best-fit curves are plotted and the equations given at the bottom.

7.7.   Conclusions

Clay distribution in a waterlogged soil valley bottom can be analysed using comparative non-linear regression method, allowing choosing the most appropriate subsurface drainage technique. Our results show that, in the case of the irrigated rice soil of Tiefora in Tropical Savannah, a quadratic curve is the most suitable to describe clay distribution (shape and location) in the soil. No unique equation could be found for the entire valley. It appeared that a soil layer made of 20-30% of clay and a thickness of 2 m persists in the topsoils from upstream to the middle areas. Downstream, where the permeability is higher, this clay content reaches a peak of 40% but with a much lower thickness of 0.50 m. In practice, these two configurations call for completely different subsurface drainage techniques. From upstream to the middle areas of the valley bottom, mole drainage and/or cracks making ditches will improve the permeability. In the downstream soils, a classical Hooghoudt perforated pipe drainage technique is appropriate to alleviate waterlogging.

8. INFILTRATION RATE INCREASE FROM UPSTREAM IN A VALLEY

Given the crucial role of valley bottom in ensuring food security, soils in these areas have been submitted to an intensive research and development (Oosterbaan et al. 1986, Nguu et al. 1988). In fact, valley bottoms – where the most sophisticated and rich traditional farming systems are found in Africa – represent less than 10% of the total crop production areas (Oosterbaan et al. 1986). However, valley bottoms constitute privileged zones for various physicochemical processes such as sedimentation, and precipitation of toxic and non-toxic ions. For example, poor drainage capacity or very low permeability is closely linked to the increase of acidity and rice iron toxicity (Chérif et al. 2009). Much research addressed the permeability characterization issue in valley bottoms. For this purpose, regional cartographic mapping of soil permeability and sensitivity were conducted (Sharma et al. 1987, Cam et al. 1996). Attempts were also made to assess the soil infiltration rate using remote sensing in arid regions (Ben-dor et al., 2004). In addition, regression fittings have been applied to describe infiltration in Savannah valley bottom soils (Fagbamia and Ajayia 1990). Though all these studies cover more or less wide zones, none of them examines the relationship between the under phreatic level permeability and the toposequence at which it is located. Nevertheless, discriminating permeability values throughout a valley, where a subsurface drainage project is planned, can lead to a drastic reduction of the implementation costs, and a much greater efficiency. The present study contributes in this area by making a comparative regression analysis of three cross sections of the valley of Tiefora in Burkina Faso (West Africa), from upstream to downstream (Keïta et al. 2014c).

8.1. Boreholes preparation

In order to get as much as possible a complete picture of the permeability in the valley, a topographic survey was conducted and nine borehole locations determined. The survey provided map of the area with the contour lines, and the indication of the limits of the irrigation scheme. The main axis and, perpendicularly to it, the upstream, the middle and the downstream cross-sections of the valley were identified. Subsequently, on each of the cross sections, three points – one along the axis of the valley and two at the edges of the cross section but inside the irrigated area – were marked and their geographic coordinates determined by topographic survey and re-checked with a Global Positioning System device (Figure 8.1).

In the 50-year old valley bottom, anticipating a heterogeneous soil requiring several boreholes, two main types of boring machines were prepared. When passing through a clayey and cohesive but non-saturated soil, the Edelman auger of 70 mm of diameter was used. The body of this auger is made of two blades that tapered into two linked "spoons" at the bottom, while they are connected at the top via a bracket to an extension rod operated by a handle (Royal Eijkelkamp 2009). Across saturated soils under the groundwater table, in layers with sand or gravelly non-cohesive material, the Riverside auger of 70 mm diameter was used. The body of this auger consists of an open tube at the bottom of which is soldered two spoon-shaped pieces. Here also a bracket connects the top of the open tube to the bottom part of the extension rod operated by a handle. Extension rods of one meter length with coupling sleeves were connected to the auger at the bottom, and were operated on the top using a screwable handle (Royal Eijkelkamp 2009). The boring would be ended any time a clear fine sand layer was found. The fine sand layer would create a higher infiltration rate at the bottom – not reflecting the soil top layers permeability – rapidly filling or emptying the borehole. Therefore, the borehole was clogged back with clay until the complete covering of that layer. These non-lined boreholes - with depths ranging from 1 to 5 m - were used for under phreatic level permeability measurements.

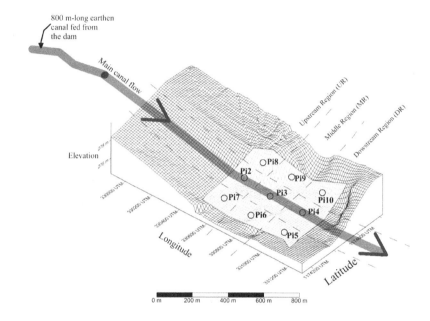

Figure 8.1: The three cross sections and the location of the nine boreholes of Tiefora valley bottom

The main canal receives water from a "transport" earthen canal, which is fed by an earthen dam created reservoir located 800 m upstream. The circles represent the piezometers (boreholes) realized along three cross-sections in the Tiefora valley bottom (Burkina Faso). Pi2 stands for "Piezometer N°2", Pi3 is the Piezometer N°3 etc. The cross sections are located some 300 m from each other. Source: (Keïta et al. 2013a).

8.2. Under phreatic infiltration rate measurements

Though several permeability measurement methods exist (Bouwer and Jackson 1974, Peter 1990) only two of them – allowing under phreatic level permeability measurements – were planned: the Hooghoudt method and the Lefranc method. The Hooghoudt method (Schilfgaarde, 1957) was not applicable because the impervious layer, supposed to be at a depth of 2 m, was simply absent. Conversely, the first boreholes revealed the existence of an impervious layer on the top of a pervious one. Consequently the Lefranc method (Cassan 2005) was used. Referring to an initial phreatic level in a soil, this method is implemented by producing a differential hydraulic head inside a cavity of known dimensions. The cavity is created at the basis of a borehole. This way, the flow towards the borehole, generated by the hydraulic head, happens only through the cavity (Figure 8.2). The evolution of the differential hydraulic head – created by pumping or injecting water at a constant discharge – was measured against the time at observable increase (or decrease) of the water level. The resulting times t and hydraulic head Hi intervals were used to compute the infiltration rate per time interval as it follows in Eq. (8.1):

$$I_t = -\frac{Hi_{t+\Delta t} - Hi_t}{\Delta t}$$
(8.1)

where: I_t is the infiltration rate (cm h^{-1}); $Hi_{t+\Delta t}$ represents the hydraulic head at instant $t + \Delta t$ (cm), and the Hi_t the corresponding value at instant t (Figure 8.2). The negative symbol is used in the current pumping Lefranc test (instead of injecting) from the borehole ($Hi_{t+\Delta t}$ smaller than Hi_t). Statistical analysis was applied to the resulting data.

8.3.   Statistical analysis

The statistical analysis strove, first, to determine two suitable regression best fit models, and to select the one with the most meaningful physical parameters. For this purpose, two non-linear regression models were tested: the quadratic and the exponential one-phase decay equations. The first selection criterion between the two models was their goodness of fit, and the second their simplicity when both displayed equivalent first criterion. Finally, the last criterion was the physical meaningfulness of the model parameters. The considerable time requirement for digging every borehole, and to get the stabilization of the hydraulic head (Figure 8.2), restricted the amount of data collected per borehole to generally less than ten. Nevertheless, the goodness of the model was assessed using three sets of key parameters: i) the goodness of fit (especially the coefficient of determination R² and the standard deviation of the residuals Syx) (Montgomery and Runger 2011); ii) the normality test of residuals of Shapiro-Wilk (Shapiro and Wilk 1965), and iii) the assessment of the deviation of the data from the curves by Runs test method (McWilliams 1990).

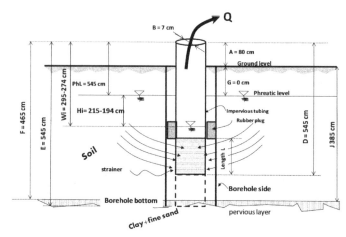

Figure 8.2: Lefranc test implemented at Tiefora

The Lefranc test for under phreatic permeability measurement is explained here using the case of the piezometer called Pi8. The equipment was installed using rubber plugs between the plastic tubing of the borehole and the impervious PVC tubing that holds the strainer at its extremity. This system composed of PVC tubing and rubber plugs forces the water – coming from the soil layers which permeability was measured – to flow into the strainer. Once the tubing prepared, a borehole 80-85 mm of diameter is implemented using 70 mm diameter augers till just before the fine sand layer. The tubing is introduced inside the borehole and the following parameters are either measured or computed. Measured parameters: coordinates (X, Y) of the borehole, tube length above ground A, tube diameter B, distance from the tube start to the strainer bottom edge D, length of the strainer L, depth of the pervious layer J, and the water level Wi depending on the time. Computed parameters: phreatic level PhL, borehole edge depth E, pervious layer start F, borehole section S, phreatic depth G from the soil surface, and the hydraulic head Hi.

Once the most suitable models were selected, the analysis focused on comparing the infiltration processes on a cross sectional basis. The major surface flow – governing sediment transports and depositions (Selley 2000) during flooding period – suggested dividing the valley bottom into three cross sections for permeability analysis: upstream, middle and downstream. Accordingly, the complete set of regression models found for the three boreholes of a considered cross section are statistically compared. This comparison checks whether the infiltration processes were equivalent or a unique permeability can be chosen for the cross section. This was done by hypothesis testing of variance of the regression parameters (Härdle and Mammen 1993, Neumeyer and Dette 2003). When the comparison failed for the entire set of parameters, the models were once more compared on the only basis of the most meaningful parameter: the permeability.

8.4.   Selecting a regression model

In spite of the goodness of fit of both quadratic and exponential one-phase decay models, they could be discriminated based on physical meaningfulness of their parameters. For example, both models fitted with a coefficient of determination R^2 greater than 0.70 (Rumsey 2009). In fact, for each of the nine boreholes, both quadratic and one-phase decay models had the same degree of freedom and therefore the p-value allowing the rejection of one of the two hypotheses could not be computed. However, while the coefficients in the quadratic Eq. (8.2) had no real physical meaning, the exponential one-phase decay Eq. (8.3) does. The two equations are:

$$I_t = B_0 + B_1 t + B_2 t^2 \tag{8.2}$$

$$I_t = \left(Y_0 - Plateau\right)\exp\left(-Kt\right) + Plateau \tag{8.3}$$

where: B_0 in cm/hours (cm/h), B_1 in cm/h², B_2 in cm/h³ are the coefficients in the quadratic equation, and t, the time variable expressed h; the infiltration rate I_t in cmh⁻¹ at instant t; the coefficients Y_0 and Plateau in cm/h; the coefficient K in h⁻¹ and the instant t in h. Plateau represents the limit of the function I_t when the time t becomes infinite.

(A)   (B)

Figure 8.3: Measuring water levels in Lefranc test

(A) Prof. Yacouba inspecting the equipment for underphreatic level test. (B) After pumping, the float rising in the piezometer Mr. Keita reads on a tape placed on top of the pipe. The time is also measured simultaneously with a chronometer.

At most, the quadratic function admits one extremum corresponding to the point where the first derivative is null. There is no limit to this function when the time t tends to infinite, and the zero intercept equal B_0 possesses no physical meaning as far as groundwater flow is concerned. By contrast, the exponential one-phase decay function was successfully used in many cases. It is applicable anytime the rate, at which one variable is altered, depends on the remaining amount of that same variable. Well-known examples include already the soil infiltration rate determination (Horton 1941), the decay of radioactive isotope atoms (Gardner et al. 1959, Varani et al. 1990), the turnover of blood cells in HiV-1 infection (Ho et al. 1995). In other words, it is most appropriate in the present case where there is a decrease in time (deceleration) of the flow rate dI / dt at a velocity that can be considered proportional to the infiltration rate I. In these conditions, the equation can be expressed as:

$$\frac{dI}{dt} = -K.I \tag{8.4}$$

where: dI/dt (cm h^{-2}) is the average flow rate change (deceleration) in the soil medium during a short duration dt; t in his the time; K (h^{-1}) is a constant of proportionality, and I (cm h^{-1}) is the infiltration rate.

It can be verified that the following improper integral is a solution of Eq. (8.4):

$$I = C_1 \exp(-K \cdot t) + C_2 \tag{8.5}$$

where: I (cm hr^{-1}) is the infiltration rate and C_1 (cm hr^{-1}), the "span" of variation of the infiltration rate with t increasing from zero to infinite (it corresponds to Y_0-Plateau in Eq. (8.3). The coefficient C_2 (cm hr^{-1}) is the "plateau", the limit of the function when t increases infinitely; and K (hr^{-1}) is the constant of proportionality.

All the parameters of Eq. (8.3) – which is the same as Eq.(8.5) – possess a clear physical meaning in the groundwater flow. That explains why this model was selected. At the beginning of the measurements in the saturated soil, when $t = 0$ the hydraulic head and hence the infiltration rate over the first minutes are the highest, the initial infiltration rate equals "span+plateau" (C_1+C_2). Then, while the experiment progresses, the hydraulic head decreases, reducing the infiltration rate. The infiltration rate depends on the remaining water layer and the process can therefore be described by an exponential one-phase decay regression. At greater values of the time t, the infiltration rate is smaller. Theoretically, when t tends to infinite, the asymptote of the function provides the value of the under-phreatic level permeability, *plateau* expressed in Eq.(8.3). Therefore, the value of the *plateau* can be used as the key parameter for subsurface drainage if any solute (e.g. excess of aluminium, ferrous iron, sulphur etc.) has to be removed from the submerged soil (Suryadi 1996).

8.5. Regression goodness of fit checks

Although many measurements points could not be obtained, the one-phase decay fitting to Tiefora permeability data provided rather good results. As it can be observed from Table 8.1, the coefficient of determination R^2 for example ranged from 0.70 to 1.00. But as one knows, R^2 is not sufficient to conclude on the goodness of fit of a model (Boslaugh and Watters 2008). The normality test of the residuals gave p-values also greater than the significance level $\alpha = 5\%$ (except for boreholes Pi3 and Pi5), allowing to conclude as for the normality of the residuals. Finally, the results yielded also an adequate distribution of the data around the curves, as shown by the values of the Runs test and Figure 8.4 (Härdle and Mammen 1993).

8.6. Cross sectional permeability comparisons

The one phase decay regression modelling from upstream to downstream revealed not only a great variability in the infiltration processes, but also in the value of the stabilized permeability or *plateau* (Figure 8.4). The attempt to make three regression equations with one related to each of the three cross sections (upstream, middle and downstream), led to the rejection of this null hypothesis of similar processes with respective p-value of 0.01% for upstream, 0.09% and 0.01% for middle and downstream respectively. Therefore for all the cross sections the coefficients Y_0, K, and *plateau* of Eq.(8.3) were different and hence the infiltration processes dissimilar.

Table 8.1: Goodness of the models

Piezo	Pi7	Pi2	Pi8	Pi6	Pi3	Pi9	Pi5	Pi4	Pi10
Selected regression type	expntl	expntl	expntl	expntl	expntl	expntl	expntl	expntl	expntl
Goodness of fit[a]									
Degree fo Freedom	1	4	2.00	2	5	2	3	4	3
Regression coefficient R^2	0.97	0.93	0.99	0.90	0.73	0.83	1.00	0.84	0.98
Absolute Sum of Squares (Ab SS)	0.00	0.06	0.18	5.03	6.05	2.15	0.23	5019.00	153.40
Standard deviation of residua	0.05	0.13	0.30	1.59	1.10	1.04	0.28	35.42	7.15
Normality test of residuals									
W of Shapiro-Wilk p-value	51%	47%	48%	18%	3%	59%	4%	36%	63%
Runs test									
Points above the curve	2	3	2	2	4	3	2	3	4
Points below the curve	2	4	3	3	4	2	4	4	2
p-value (runs test)	100%	54%	90%	90%	63%	90%	80%	97%	100%
Deviation from the model	Not significant	Not significant	Not significant	Not significant	Not significant	Not significant	Not significant	Not significant	Not significant

[a]Null hypothesis = the most suitable regression equation is exponential one-phase decay (expntl);
Alternative hypothesis =the best fit function is a quadratic (quad) function (second order polynomial).

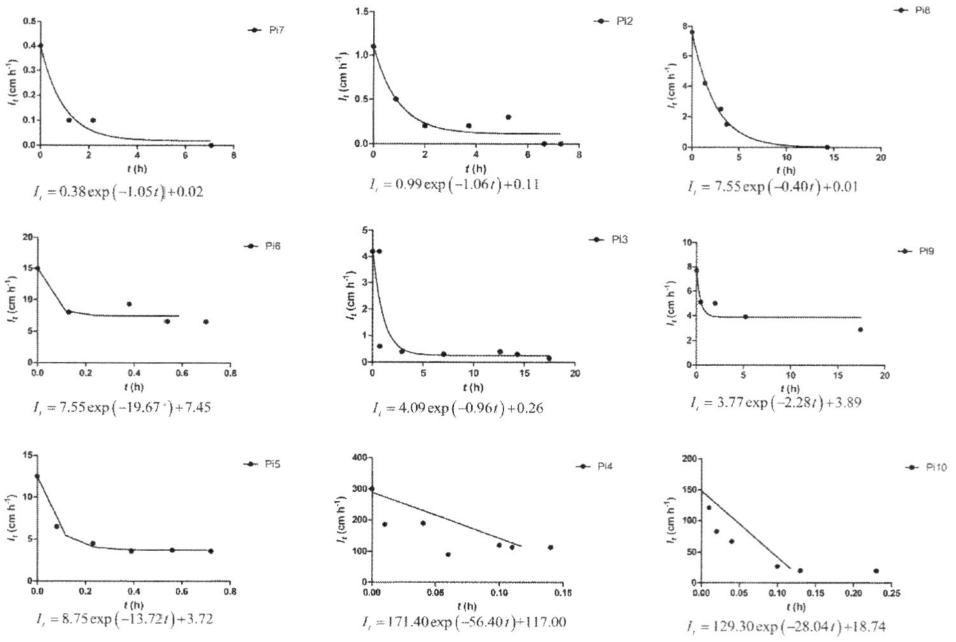

Figure 8.4: Regression best fit curves and equations of the boreholes

The three piezometric boreholes Pi7 Pi2 and Pi8 data on top represent the upstream cross section; the data of Pi6, Pi3 and Pi9 the middle cross section, and the data of Pi5, Pi4 and Pi10 the downstream cross section. The exponential one-phase decay regression curves represent the individual best fit curves, when no comparison is made among the boreholes infiltration processes.

The important variability in the permeability – not very convenient for drainage engineering – can be lessened on valid statistical basis. In fact, subsurface drainage design uses a small number of permeability values to implement projects in swampy areas (Wesström et al. 2001, Keïta et al. 2013b). Hence, one can loosen the constraints upon the infiltration processes and statistically compare only the *plateau* values in each cross section. This examination reveals that the under phreatic level permeability is similar (p-value greater 5%) for the three boreholes (Pi7, Pi2 and Pi8) in the upstream cross section. Their *plateau* equals 0.10 ± 0.10 cm/hr. They are also similar for the three boreholes (Pi6, Pi3 and Pi9) in the middle section, where the *plateau* equals 0.32 ± 1.30 cm/h(Table 8.2). Therefore, though the infiltration processes are different, a single permeability can be used for the upstream cross section on one hand, and the middle cross section on the other hand. However, the variability is so important in the downstream cross section that a unique permeability cannot be chosen.

The remarkable increase of the permeability from upstream to downstream in the valley bottom (Figure 8.1 and Table 8.2) cannot simply be explained by soil alteration. Two major phenomena are generally considered to explain valley bottoms genesis (Schaetzl and Sharon 2005). First, it is assumed that valley bottoms and streams generally appear on the top of deep fractures of the sound rocks (Comité Inter-Africain d'Etudes Hydrauliques (CIEH) 1987) as the consequence of the rock chemical alteration driven by acidic water (Jackson and Sherman 1953). Then, the fracture develops transversal and longitudinal slopes and clay tends to accumulate. The second phenomenon, simultaneous to the first, consists of periodic or continual deposits of thin layers of colluvial, alluvial or eolian sediments (Moormann and Breemen 1978). That deposit changes the profile. These theories do not address the change in permeability from upstream to downstream.

From a different standpoint, the sedimentation process can greatly help explaining the toposequential increase of the permeability. The sediments transport and deposition processes have been object of several important studies (Selley 2000). As shown by the results (Table 8.2), the permeability is much higher downstream with 20.00 ± 10.00 cm/h than upstream with 0.10 ± 0.10 cm/hr. This observation can be explained to some extent by heterogeneity downstream. The soil data showed a thicker clay layer upstream and in the middle than downstream, where a quartzic stony soil is found much closer near the soil surface. However, in lakes and reservoirs, the water flow conveys water from a narrow irregular channel to an extended area, causing the drop of velocity. Heavier particles like sand deposit first, while fine clay will deposit in the reservoir bed later but in greater amount near the dam located farther. The permeability would therefore be lower near the dam. The phenomenon is inverted in flood plains: when the sediment loaded water invades the area, fine clay particles are deposited farther, upstream. Therefore, the permeability will be smaller upstream. We think this last process applies to the case of Tiefora where the valley bottom continues to be flooded from downstream twice a year by the Lafonde river (Kanté 2011).

The practical implications of such a toposequential variation of the permeability are important. In effect, it is well known (Skaggs 2007) that the spacing of subsurface drainage pipes is smaller when the under phreatic permeability is small, in order to be able to intercept more important flow. Therefore, drain pipe density in a valley bottom with significant permeability increase from upstream to downstream will also conversely decrease from upstream to downstream. This consideration can lead to a more effective drainage at lower cost in comparison to a uniform piping of the swampy area.

Subsurface drainage of valley bottom in Tropical Savannah

Table 8.2: Regression parameters with 90% confidence interval

Piezo	Pi7	Pi2	Pi8	Pi6	Pi3	Pi9	Pi5	Pi4	Pi10
		Upstream Section			Middle Section			Dowstream Section	
	expntl	expntl	expntl	expntl	expntl	expntl	expntl	expntl	expntl
Selected regression type									
Y0	0.40 ± 0.31	1.09 ± 0.30	7.56 ± 0.30	13.02 ± 2.29	4.35 ± 2.78	6.16 ± 1.74	12.47 ± 0.65	288.5 ± 73.20	48.10 ± 33.3
K	1.94 ± 8.53	0.93 ± 0.13	0.40 ± 0.72	1.21 ± 0.62	0.99 ± 1.73	0.06 ± 0.08	13.72 ± 3.12	56.4 ± 78.80	28.04 ± 15.74
Span (=Y0-Plateau)	0.32 ± 0.33	1.02 ± 0.29	7.50 ± 0.32	12.70 ± 2.61	4.30 ± 3.00	5.84 ± 2.14	8.75 ± 0.73	171.40 ± 79.5	129.30 ± 30.7
Plateau	0.07 ± 0.13	0.07 ± 0.13	0.07 ± 0.13	0.32 ± 1.30	0.32 ± 1.30	0.32 ± 1.30	3.72 ± 0.36	117.00 ± 40.80	18.74 ± 12.95
Goodness of the models									
R square	0.93	0.92	0.99	0.68	0.73	0.68	-1.18E-9	0.70	0.97
Absolute Sum of Square	0.01	0.07	0.18	15.55	6.07	4.17	61.13	9305.00	246.70
Standard deviation of residuals (Sy,x)									
Unique plateau p-value for the cross section	84.40%				5.20%			Individual plateau values for each borehole	
Selected unique plateau va	0.07 ± 0.13				0.32 ± 1.30		3.72[a] ± 0.36	117.00 ± 40.80	8.74 ± 12.95

high value of *plateau* is due to quartzic stony layer found in the Pi4 borehole. [a]This

8.7.   Conclusions

Three main conclusions were reached. First, comparative regression analysis can be used to assess the equivalence of infiltration processes, and to determine a toposequential distribution in a valley bottom with very positive economical implications for subsurface drainage. Secondly, this analysis was applied to the Tropical Savannah irrigated rice valley of Tiefora in Burkina Faso (West Africa). Our results show that no unique infiltration process is valid for the whole valley. However, confining the comparison criterion to the only under phreatic permeability reveals that it increases for upstream with 0.10 ± 0.10 cm/hto more than 20.00 ± 10.00 cm/hin some areas downstream, in spite of the differences of the infiltration curves. Finally it appeared that, taking into account that subsurface drainage pipe spacing is bigger – and then the density of implantation is smaller – when the permeability increases, a greater efficiency and financial saving in drainage system design and operation of valley bottoms can be achieved.

9. WATER MANAGEMENT USING AUTOCORRELATION

The fast growing global population and rural poverty put efficient water management in irrigation systems as a central challenge. In the coming 50 years, many countries in Asia, Africa and Middle East are expected to double their populations. In parallel, the food production will have to double to ensure a global food security (Schultz et al. 2009). As illustrated by the green revolution during which over 70% of the food increase were provided by irrigated lands (Seckler 1996), the expectations about irrigation systems are great. Irrigation is known as one of the most important rural poverty alleviating factor (Hussain and Hanjra 2003, Namara et al. 2009). But, developing new infrastructures faces increasing environmental, social and economic constraints (Nguyen and Ferrero 2006). While the crop production is expected to increase, good quality irrigation water will decrease because of the growing competitive water demand of industries and municipalities (Oster 1994). Therefore, irrigation is confronted the challenge to produce more food with less water. In addition, excessive water use in irrigation may result in poor drainage conditions and the degradation of soil quality with the subsequent metal toxicities for crop production (Achim and Thomas 2000). Hence, water management in existing irrigation schemes is critical.

While accused of being the most important inefficient irrigation systems in use in Africa, surface irrigation strangely also suffers from a lack of follow up and analysis of water use. Surface irrigation represents the first irrigation technique in West Africa (Frenken 2005), covering more than 2 million ha (99% of the total irrigated area). Many of the surface irrigation schemes produce rice in valley bottoms, known to be one of the biggest water consumers (Becker and Johnson 1999). When there is poor water management, not only water that could be used to extend the irrigated area is wasted, but also the inherent non-equity in water distribution generates frictions among farmers (Pretty JN et al. 2003). In addition, soil degradations such as iron or aluminium toxicities often result from poor drainage conditions (Becker and Asch 2005). Surprisingly, one could not find in the available research on irrigation in West Africa any statistical modelling analysis of time series records (Montgomery et al. 2008) related to water use in surface irrigation systems. The current study addressed this issue through the case of the 16 ha rice irrigated valley bottom of Tiefora in Burkina Faso. It applied the trend analysis and autocorrelation modelling – particularly the ARIMA model (Shamway and Stoffer 2011) – to assess the daily water use during the irrigation season and evaluate farmers' water distribution schedule (Keïta et al. 2014b).

Figure 9.1: Tiefora rice irrigated valley environment

On can see the 800 m long earthen transport canal conveying water from the earthen dam formed reservoir to the irrigated valley bottom. The water level diver is installed at the beginning of the main irrigation canal, some 13 meters upstream of a concrete weir.

9.1.   Water level diver installation

In order to follow up as accurately as possible the irrigation water use in the 16 ha irrigated rice scheme of Tiefora, it was decided to install an e-water level diver (Schlumberger 2006) at the beginning of the main canal. The diver was installed (Keïta 2013a) before the first secondary canal offtake, and some 13 m upstream from a broad crested weir whose height is 0.38 m (Figure 9.1 and Figure 9.2). The installation of the diver some 3-4 m upstream the weir as usually recommended (Henderson 1966, Chaudhry 2008) would have been enough to ensure a normal depth at the diver location, but a wide flagstone crossing the canal prevented using such a location. The purpose was to use the flow equation of the weir to compute the flow rate introduced in the canal. Fluctuations of the water level were expected for various reasons: an important demand of water by farmers having missed their irrigation turn, farmers irrigating during the nights out of normal working period, leakage of the gate, gate opened by those fishing in the main canal etc. Hence, the diver was scheduled to do an automatic measurement of the water depth H_{meas} every 6 hours, beginning at 6:00 am. The four measurements taken per day are averaged to produce the daily water use (DWU) time series using the weir formula (Chaudhry 2008):

$$Q = CB\sqrt{2g} \times H_{meas}^{\frac{3}{2}} \tag{9.1}$$

where: Q is the discharge; C is the correction coefficient (0.40 in the concrete case of Tiefora); B is the width of the channel (0.44 m in the case of Tiefora); H_{meas} is the water depth measured by the e-water level diver.

The discharges Q_1, Q_2, Q_3, Q_4, computed according to Eq. (9.1) with the measurements $H_{meas,1}$, $H_{meas,2}$, $H_{meas,3}$, $H_{meas,4}$, are multiplied each one by 6 hours to get the corresponding water volumes. Hence, the water volumes are added to get the daily water use (DWU) in m^3/d. The so obtained time series were recorded during 467 days, i.e. three irrigation seasons of 4 months in Tiefora: one rainy season and two dry seasons (no rainfall occurs during the dry season).

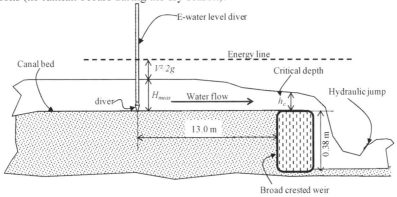

Figure 9.2: E-water level diver installed on the beginning of the main canal

The diver was installed at the beginning of the irrigation main canal, before any secondary canal offtake. It was programmed to measure the water depth every 6 hours.

9.2.   Assessment of irrigation water use during the seasons

To assess the irrigation schedule in the valley bottom of Tiefora, two datasets had to be compared. The first dataset is obtained through field survey. Questions were asked to a sample of 6 farmers in addition to the "president" of the farmers organisation about how irrigation is organized. These investigations led to the production of the irrigation cycle in Tiefora, for both the rainy season – during which it can rain up to 1200 mm – and the dry season – during which there is generally no rainfall. The farmers' effort to take into account the rain water availability to reduce the irrigation water use was also appreciated. The second dataset was drawn from the daily water use (DWU) records measured by the e-water level device. To get the evolution of the irrigation schedule along the timeline, the classical technique (Moreau and Pyatt 1970, Vieux and Farajalla 1994) of moving average (MA) smoothing was applied to the highly fluctuating data records. By comparing these two datasets, an evaluation of the irrigation schedule of the farmers' organisation was performed.

9.3.   Statistical analysis

9.3.1. Removing outliers and stabilizing the variance

The initial daily water use data included outliers making it difficult to perform sound statistical analysis (Spiegel et al. 2001). As usual, outliers affected two statistics in the sample data: the mean and the variance. Thus, taking into account the irrigation scheme size of 16 ha and the type of crop grown in the valley bottom which is rice (ETM= 70 m^3/ha/day in the region), it appears that extreme DWU high values – i.e. greater than 1500 m^3/day corresponding to the irrigation of some 4 ha in Tiefora – had to be eliminated. That was done using a boxplot (Figure 9.3). The original DWU values of Tiefora contained some 467 measurements made with the E-water level diver (Schlumberger 2006). After elimination of the outliers, the time series size decreased to 392 observations, still enough for a sound time series analysis (Box et al. 1994).

The second challenge with the data consisted in the non-constancy of the variance, which tended to expand as the time goes on. This phenomenon is quite frequent in time series (Shamway and Stoffer 2011) and several transformation techniques exist to stabilize the variance. Two of them are logarithm base 10 and square root functions applied to the original data. After fitting a time series model to the data, the goodness of fit is checked by plotting the residuals versus, for example, the orders of observations or measurements. This plot must display a cloud of points with a constant width around zero (Boslaugh and Watters 2008). The square root sqrt(DWU) was the most successful with the data of Tiefora because it allowed to later obtain a model with stabilised residuals. Once these pre-processing had been done, modelling techniques were applied to the data as explained in the next sections.

9.3.2. ARIMA model

It is important to recall that the daily water use data of Tiefora are a "time series" and the implications of such type of data are multiple. Very often, time series data are not independent and are not distributed in an identical manner: a future observation depends on the current and the past observations, i.e. adjacent observations are dependent (Box et al. 1994). There is autocorrelation in the data. In fact, when the 800 m long irrigation main canal of Tiefora is full of water today, it is more likely that it will also have some water tomorrow for the crops do not instantly disappear from the field. On the contrary, if the main canal is empty today, there is a great chance that it will also be empty tomorrow. This persistence or tendency to remain in the same state often characterises geophysical systems. The underlying autocorrelation has both downside and upside. On the downside, it illegitimates most of the statistical hypothesis tests which require that the observations are independent and identically distributed (constancy of variance and mean) (Levene 1960, Brase and Brase 2007, Rumsey 2009). On the upside, autocorrelation provides means to predict future observations from past historical records (Miller 2013).

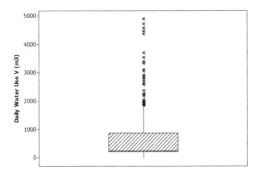

Figure 9.3: Identification of the outliers in the time series DWU

The "*" represent the outliers.

After careful examination of the DWU (or x_t) time series data of Tiefora, the Autoregressive Integrated Moving Average (ARIMA) or Box & Jenkins model was judged suitable. ARIMA is a model that can efficiently reproduce a complex process including four properties (Tsay 2005). The first is an auto regression AR, i.e. a process showing the linear dependency of the current observation x_t on the h past observations $x_{t-1}, x_{t-2}, .., x_{t-h}$ within the time series itself. The second property is a moving average MA, i.e. a process showing the linear persistence in the current observation x_t of random shocks or measurement errors $w_{t-1}, w_{t-2}, .., w_{t-h}$ done during the previous h observations which add up to an average value μ (Montgomery et al. 2008). The third property consists of the existence of a trend D or long-term movement (increase, decrease or stagnation) in the time series x_t. Finally, the fourth property is the presence of seasonality of span S, i.e. the existence of a specific pattern in the time series x_t that repeats identically after a number of time units equal to S (Agung 2009). By definition, one would expect that the seasonal span S in the irrigation data record DWU would correspond to the repeating water distribution pattern time span T set by the farmers to deliver equitably water to those farms within the so called tertiary unit (de Veer et al. 1993). This time span is also called "tour d'eau" (Dembele et al. 2012). The pre-mentioned four properties are often found in chronological records of flow rates of rivers and thus (Salas et al. 1980), were expected to be found the DWU records of Tiefora. Therefore, appropriate components AR, D, MA and S were searched thoroughly by iteration using the statistical package Minitab (Montgomery 2001, Rumsey 2009) in order to determine the suitable ARIMA model to apply to the data of Tiefora.

A successful modelling with ARIMA proceeds from understanding how the equations are constructed. ARIMA uses two key operations to rebuild non-stationary time series processes. These two operations are backshift and backward operations (the last is associated with differencing to remove trends and seasonality in the data). These operations are defined in equations. There are mainly two equation forms of the seasonal ARIMA model. The first form possesses the advantage to incorporate both non-seasonal and seasonal factors in a multiplicative model and can be written (Shamway and Stoffer 2011):

$$ARIMA(p,d,q) \times (P,D,Q)_S \qquad\qquad (9.2)$$

The second form of the model shows more explicitly the operations involved and is written as it follows (Montgomery et al. 2008):

$$\Phi(B^S)\phi(B)(1-B)^d(1-B^S)^D x_t = \Theta(B^S)\theta(B)w_t \quad (9.3)$$

where there are:

x_t = the time series

w_t = white noise (pure random process, without any autocorrelation in it, with 0 as mean and σ_w^2 as variance)

Non-seasonal factors

ϕ = autoregressive (AR) polynomial coefficient operator defined by:

$$\phi(B) = 1 - \phi_1 B - \phi_2 B^2 - \ldots - \phi_p B^p ;$$

where ϕ_i are coefficients.

θ = moving average (MA) term coefficient operator defined by:

$$\theta(B) = 1 + \theta_1 B + \theta_2 B^2 + \ldots + \theta_q B^q ;$$

where θ_i are coefficients.

p = non-seasonal autoregressive component order

d = non-seasonal differencing order

q = non-seasonal moving average component order

B = backshift operator (move x_t back one step: $Bx_t = x_{t-1}$)

$(1-B)^d = \nabla^d$ = backward or differencing operator applied d times. This operator is defined by : $\nabla x_t = x_t - x_{t-1}$

Seasonal factors

Φ = autoregressive (AR) polynomial coefficient operator defined by:

$$\Phi(B) = 1 - \Phi_1 B - \Phi_2 B^2 - \ldots - \Phi_p B^p ;$$

where Φ_i are coefficients.

Θ = moving average (*MA*) term coefficient operator defined by:

$$\Theta(B) = 1 + \Theta_1 B + \Theta_2 B^2 + \ldots + \Theta_Q B^Q ;$$

where Θ_i are coefficients.

P = seasonal autoregressive component order

D = seasonal differencing order

Q = seasonal moving average component order

S = season time span

B^S = backshift operator applied S times (moves x_t back S steps : $B^s(x_t)=x_{t-S}$)

$(1-B)^D$ = backward operation applied D times

The construction methodology of the model application to the specific dataset of Tiefora is based on the utilisation of two tools defined below: the autocorrelation function (ACF) and the partial autocorrelation function (PACF) (Tsay 2005).

9.3.3. Autocorrelation

The so-called autocorrelation function (ACF) possesses at least two great interests. The first–that follows from its definition – lies in the fact that its values become zero or non-significant after lag q for a stationary moving average process (MA) of order q (q being the number of previous observation errors w_t to use to predict the current observation x_t) (Hamilton 1994). This property – clearly visible on the ACF plot with the coefficient bars dying out after few lags – was applied to the DWU process of Tiefora in order to see whether it is a pure moving average. The autocorrelation coefficient ρ_h between two observations x_t and x_{t-h} – made h time-units apart – measures the strength of their correlation within the time series. As described by several authors (Montgomery et al. 2008, Miller 2013), this autocorrelation coefficient at lag h for a weak stationary (constancy of variance and of mean over timeline) time series is defined by the following ratio:

$$\rho_h = \frac{\text{Covariance}(x_t, x_{t-h})}{\text{StdDev}(x_t) \cdot \text{StdDev}(x_{t-h})} = \frac{\text{Covariance}(x_t, x_{t-h})}{\text{Variance}(x_t)} \quad (9.4)$$

where: ρ_h = autocorrelation coefficient at lag h; t = time expressed in calendar format or as an integer used as index ; x_t = the value of the time series at time t ; h= 1, 2, 3, etc. represents the time lag index. The collection of ρ_h constitutes the autocorrelation function, ACF.

The second interest of ACF resides in the fact that – even if the process is neither stationary nor pure moving average – its plot can be used to determine the MA orders to include in a more general non-stationary (existence of trend and seasonality in the data) ARIMA model (Box et al. 1994). In that case, if the ACF applied to the data presents sharp spikes at small h lags, that suggests to include an MA of order q term in the non-seasonal component (q is the lag corresponding to the last early significant spike in the plot) in the ARIMA. In addition, wavelike bars in the plot with spikes at h lags that are multiple of the seasonal span S of the time series suggest to include an MA term of order Q (Q being the number of previous seasonal observation errors w_{t-S} to use to predict the current observation x_t) in the seasonal component of the ARIMA (Tsay 2005, Shamway and Stoffer 2011). The 5% significant limit lines at lag h for ACF, when two tailed test is used, are given by the following expression (Anderson 1941, Salas et al. 1980):

$$r_h(95\%) = \frac{-1 \pm 1.96\sqrt{N-h-1}}{N-h} \quad (9.5)$$

where : r_h = is the ACF confidence limit value at lag h, N= sample size; t = time expressed in calendar format or as an integer used as index ; h= 1, 2, 3, etc. represents the time lag index.

The autocorrelation function (ACF) was used to determine whether to include or not a moving average term MA in the ARIMA (Box et al. 1994).

9.3.4. Partial autocorrelation PACF

The so-called partial autocorrelation function (PACF) also presents two important properties. The first is that for a pure stationary (constancy of variance and mean of the data) autoregressive process of order p (p being the number of previous observations to use to predict the current one x_t), the PACF coefficients become null or non-significant after lag p (Box et al. 1994, Agung 2009). This property – clearly visible in the plot of PACF – was used with the daily water use DWU data in Tiefora to determine whether it is a pure autoregressive (AR) process. The partial autocorrelation coefficient ϕ_h between two observations x_t and x_{t-h} – made h time-units apart – measures the strength of their correlation by subtracting the linear contributions to x_t and x_{t-h} of all the set of h observations – namely x_{t-1}, x_{t-2}, ..., x_{t-h+1} – that come in between them. Hence, it is a conditional correlation (Salas et al. 1980, Shamway and Stoffer 2011). As given by several authors (Salas et al. 1980, Tsay 2005), the partial autocorrelation coefficient at lag h for a weak stationary time series is the following ratio:

$$\phi_h = \frac{Covariance\left(x_t, x_{t-h} \middle| x_{t-1}, x_{t-2}, ... x_{t-h+1}\right)}{StdDev\left(x_t \middle| x_{t-1}, x_{t-2}, ... x_{t-h+1}\right) \cdot StdDev\left(x_{t-h} \middle| x_{t-1}, x_{t-2}, ... x_{t-h+1}\right)}$$

$$= \frac{Covariance\left(x_t, x_{t-h} \middle| x_{t-1}, x_{t-2}, ... x_{t-h+1}\right)}{Variance\left(x_t\right)}$$ (9.6)

where : ϕ_h = partial autocorrelation coefficient at lag h; t = time expressed in calendar format or as an integer number used as index ; x_t = the value of the time series at time t ; h= 1, 2, 3, etc. represents the time lag index (Montgomery et al. 2008).The collection of ϕ_h is the partial autocorrelation function, PACF.

The second important property of PACF consists in the fact that – even if the process under consideration is neither stationary nor pure autoregressive – its graph can be used to find out AR coefficient orders to include in a more general non-stationary ARIMA model (Box et al. 1994). To achieve that, if the PACF applied to the data presents sharp spikes at early h lags, it is advised to include an AR of order p term in the non-seasonal component (p is the lag corresponding to the last early significant spike in the plot) in the ARIMA. In addition, wavelike bars in the plot with spikes at h lags that are multiple of the seasonal span S of the time series suggest to include an AR term of order P (P being the number of previous seasonal observation x_{t-S} to use to predict the current observation x_t) in the seasonal component of the ARIMA (Tsay 2005, Montgomery et al. 2008).

The partial autocorrelation function (PACF) was applied to check whether a moving average term MA should be taken into account in the ARIMA (Box et al. 1994) model of Tiefora.

9.4.   Assessing the water use along the two seasons

Though highly hectic (Bender and Simonovic 1994), the data displayed in Figure 9.4 show a significant moving average (MA) pattern which, compared with the farmers survey, led to the conclusion that they endeavour to take into account water availability along the season. The moving average applied here more for revealing the patterns and the trend than finding the best fit for forecasting purpose (Montgomery et al. 2008). The minimum values of the indicators MAPE, MAD and MSD (respectively 212, 224 and

88132) corresponding to the best fit among several moving average spans – namely 5, 10, 15, 20, 25 days – yielded an optimum of 20 days (Figure 9.4). It appeared that the rainy season (June- October), with less water withdrawal, generally starts in the region every year in May-June (Keïta et al. 2013b). The related moving average line indicates a reduction of *DWU*, with the lowest consumption of 200 m^3/d (Figure 9.4) which corresponds to the basic flow of the main canal (Salas et al. 1980). *DWU* is also low in January-February, corresponding to nursery period for the dry season. Furthermore, the MA indicates higher *DWU* during the 8-month long dry season (October-May) in which it normally does not rain. During that period, the *DWU* can reach a peak 1500 m^3/day. The slight drop of the MA curve, in the 2014 dry season, points to a water shortage confirmed by field investigations. This shortage is consecutive to the lack of water in the reservoir of Tiefora. Therefore, compelled either by irrigation water shortage or out of a desire of better management, the farmers' organization did adapt the *DWU* during the investigated period.

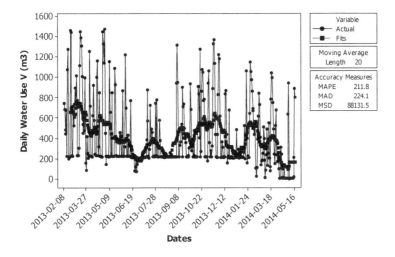

Figure 9.4: Moving average of the time series of Daily Water Use

Legend: MAPE = Mean Absolute Percentage Error; MAD = Mean Absolute Deviation; MSD = Mean Squared Deviation (Montgomery et al. 2008).
A clear irrigation pattern appears in this moving average plot. One can notice that *DWU* during the rainy season of June-September drops, while reaching a peak in January-February at the beginning of the crop growing campaign of the dry season. The general trend of the moving average line from January to May 2014 is the consequence of a water shortage confirmed by field survey.

The almost inexistence of closure of the main canal gate which is another amazing aspect revealed by the moving average plot (Figure 9.4), has numerous consequences on the rice soil. During the period from February 2013 to January 2014, the gate of the main canal was almost never closed. The discharge is barely null and almost never lower than 200 m^3/d for 4 ha (i.e. 50 mm/d vs. a local reference evapotranspiration of 7mm/d), while it should have been zero at least during the harvest period of the rainy season. This period is November-December. The gate was also expected to be closed during the rainiest months of the year, which are August and September. Therefore, though there is an effort – conscious or not – to adapt the irrigation schedule to the availability of the dam and rainfall water, it seems that the water management needs to pay more attention to water saving (Wopereis et al. 1999, Schultz et al. 2009). This almost permanent flow of irrigation water certainly contributes to water table rising, often accused as one of the main factors creating anoxic conditions in the soil, and eventually leading to iron toxicity in the region (Jacq and Ottow 1991, Chérif et al. 2009). However, as shown by Keïta et al. (2013b), during dry periods, the oxidation of hematite (Fe_2O_3) soils like found in Tiefora, though decreasing the ferrous iron (Fe^{2+}) concentration, tends simultaneously to increase soil acidity, and thus facilitates the absorption of the toxic Fe^{2+} by the rice. The permanently wet conditions found in Tiefora, although water consuming, helps in alleviating this toxicity (Keïta et al. 2013a).

9.5.   Assessing the irrigation schedule by a seasonal ARIMA model

The autoregressive moving average (ARIMA) model – constructed with the objective to find out the irrigation cycle (or "tour d'eau") – can be clearly determined using the systematic procedure previously explained. After computation of the square root of *DWU* in order to stabilize the variance of the raw data, the plots (Figure 9.5) of the autocorrelation (ACF) and of partial autocorrelation (PCAF) functions reveal the existence of adjacent dependency in the data (Box et al. 1994). As previously explained, the presence of significant spikes of ACF and PACF at early lags suggests the values of the non-seasonal components orders (Shamway and Stoffer 2011). It appears in Figure 9.5 that both ACF and PACF are significant at the first two lags. These observations lead to the orders $p = 2$ for the autoregressive and $q = 2$ for the moving average non-seasonal components of the ARIMA. In addition, no clear trend could be observed in the data plotted in Figure 9.4, which suggests a non-seasonal differencing order of $d=0$ (Hamilton 1994). The seasonal component orders P, D and Q of the ARIMA can also be drawn from the two plots in Figure 9.5. The ACF and PACF, although tapering both in a wavelike shape, present significant spikes at lag 20 and 40 days. The "crests" of the wave fade in inside the 5% significant limits (Eq(9.5)) more rapidly in the ACF – characterizing the moving average component – than in the PACF – characterizing the autoregressive component (Montgomery et al. 2008). Therefore, the seasonal component orders of the ARIMA proposed are $P = 0$ for the autoregressive part, $D = 1$ for the seasonal differencing order of span $S = 20$ days, and $Q = 1$ for the moving average part. Hence, the proposed ARIMA model written in the form of Eq(9.2) is as it follows:

$$ARIMA(p,d,q) \times (P,D,Q)_S = ARIMA(2,0,2) \times (0,1,1)_{20} \qquad (9.7)$$

The estimation of the model represented by Eq.(9.7), based on three sets of indicators show that it fits well the daily water use data of Tiefora. The three sets of indicators are: i) the Ljung-Box Chi-Square statistic for the model coefficients; ii) the normality plot of the residuals; and iii) the ACF and PACF of the residuals (Tsay 2005). The model coefficients and the Ljung-Box statistics for the residuals are presented in Table 9.1 as rendered by Minitab computations (Montgomery et al. 2008). The Ljung-Box statistics for the residuals present *p-values* all greater than 5% (i.e. the coefficients in the residuals are null or not significant), showing that the residuals of Sqrt (DWU) versus ARIMA fit are uncorrelated. That can also be checked in the ACF and PACF plots of the residuals on Figure 9.6: there is no significant value. In addition, Figure 9.7

shows that the residuals are normally (Gaussian) distributed (histogram and normality plots) and that the variance is stabilized (residuals vs. order plot). Therefore, the model in Eq(9.7) did capture all the deterministic part in the daily water use data of Tiefora and fits them well (Shamway and Stoffer 2011).

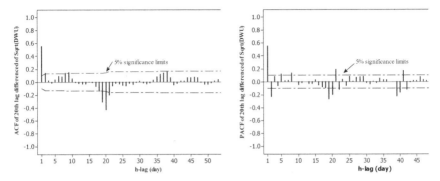

Figure 9.5: Autocorrelation Function and Partial Autocorrelation Function for $\sqrt{DWU}$

The dotted lines represent the 5% significance limits for the autocorrelations. The autocorrelation is significant at lag X if only the autocorrelation function (ACF) represented by the bar is greater than the significance limit.

Table 9.1: ARIMA model parameters for Sqrt(DWU) of Tiefora

Model parameters				
Type	Coef	SE Coef	T-stat	p-value
AR 1	1.1646	0.0916	12.71	0.0%
AR 2	-0.1699	0.0908	-1.87	6.2%
MA 1	0.4484	0.0828	5.42	0.0%
MA 2	0.4642	0.0703	6.60	0.0%
SMA 20	0.9107	0.0314	29.05	0.0%

Differencing and statistical seasonal period				
Differencing		Seasonal period	Nbr of observations	
Non seasonal d=0	Seasonal D=1	Period S = 20 days	Original series: 467	After differencing: 392

Model goodness of fit assesment parameters			

Residuals			
Sum of squares	Mean squares	Degree of freedom	
SS = 14087.1	MS = 38.4	DF = 367	

Modified Box-Pierce (Ljung-Box) Chi-Square statistic			
Lag (days)	12	24	36	48
Chi-Square	11.1	16.4	26.7	39.1
DF	7	19	31	43
p-Value	13.3%	62.9%	68.6%	64.1%

The comparison of field investigations about the irrigation cycle (tour d'eau) and the daily water use data reveal an important gap between the planned schedule and the reality. According to the farmers' schedule, the irrigation turn is 5 days: a group of farm plots of about 4 ha (Figure 9.8) receive water in day 1, a second group in day 2 etc. till day 4, but in day 5, the main canal gate is supposed to be closed because of the market event. The market of Tiefora is held every five days, and farmers sell various products during this fair. The "seasonal" pattern or irrigation cycle with a span of five days is visible on Figure 9.8. It should have been also visible in the daily water use record obtained using the ewater level device. But the ARIMA model built on the data does not confirm such a seasonal length (Moreau and Pyatt 1970). Rather, it indicates a seasonal span or irrigation cycle of 20 days (Figure 9.5), four times longer. This means that the theoretical cycle is not rigorously observed in the irrigation system of Tiefora. Equity in water distribution is achieved over a period of 20 days instead of five days. This reality is explained when one can notice that the 12 sluices of the gates in Tiefora are all hand manipulated and several of them are broken down. In such conditions, equity is difficult to achieve (Becker and Johnson 2001). To reach some equity in five days in Tiefora, the sluices and the gates will need to be fixed and a more rigorous organization in the irrigation restored (Masiyandima et al. 2003, Dembele et al. 2012). Knowing and accepting that 20 days are currently required to reach equity in water distribution may defuse potential conflicts among farmers (Pretty JN et al. 2003).

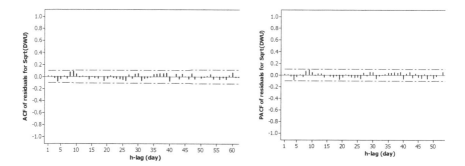

Figure 9.6: Autocorrelation Function and Partial Autocorrelation Function of residuals for $\sqrt{DWU}$

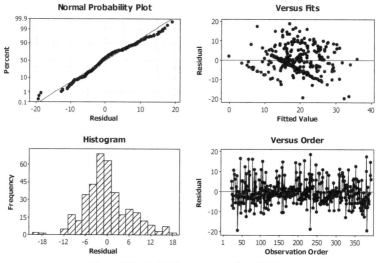

Figure 9.7: Model fit assessment Sqrt(DWU)

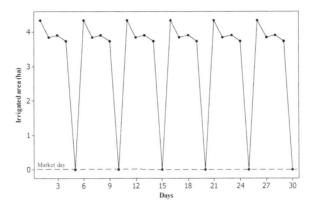

Figure 9.8: Farmers organization irrigation schedule or "tour d'eau"

The cycle is based on a period of span 5. During the period, irrigation is performed from day 1 to 4. The 5th day is held a public market and there is – in theory – no irrigation that day.

9.6.   Conclusions

Autoregressive Moving Average (ARIMA) modelling was used to assess water management of the 16 ha valley bottom irrigated rice system of Tiefora in Burkina Faso. A systematic approach, based on the examination of significant spikes of the autocorrelation and the partial autocorrelation functions applied to a time series records of daily water use yields the irrigation schedule key features. In the case of Tiefora, our results show that the main canal flow is almost never cut, with a minimum discharge of 200 m^3/d for 4 ha, while it was expected to be null every 5th day and during harvest periods. There is an excessive water use in the scheme, which calls for a better management and the rehabilitation/modernization of the gate sluices. In order to solve potential conflicts, farmers need to accept that a 20-day cycle of irrigation is required to ensure equity. Finally it appeared that installing a main water level measuring diver –at the headwork of a surface irrigation system – and producing a time series can provide, with the use of ARIMA, a thorough assessment of the water management at low cost.

10. DRAINAGE AND LIMING IMPACTS ON FEROUS IRON

Rice production faces several problems that need to be addressed as key factors to ensure food security at global scale (Dorosh 2001, Birmani et al. 2003). Rice is a global cereal demarcates from several other cereals by the fact that it is mainly auto-consumed, being internationally marketed for only 7% of the total production (Wailes 2005). It was estimated that global food production will have to be increased by about 40% at least over the next 25 years to meet the demands of a growing global population expected to reach 8.8 billion by 2050 (Lotze-Campen et al. 2008). Food security is a crucial issue in the world. Within this framework, rice is the staple food for more than three billion people, grown by more than half of the world's farmers (Fairhurst and Dobermann 2002). However, rice production faces several challenges among which one can mention the reduction of lowland areas suitable for rice due to climate change; the depletion of soil fertility; the lack of seed renewal, insects infestations, metallic ion such as zinc, aluminium and iron toxicities and soil acidification (Fairhurst and Dobermann 2002, Nguyen and Ferrero 2006, Schultz et al. 2009, Charles et al. 2010).

Iron toxicity can drop rice yield to null. It is one of the biggest challenges to rice productivity improvement that has been addressed by several fields of research (West Africa Rice Development Association (WARDA) 2006), though hydraulics contribution to solving the issue remained limited (Tanji and Kielen 2002, Ritzema et al. 2008, Keïta et al. 2014b). These fields can be dispatched into four areas of investigation: microbiology, agronomy, pedology and hydraulics. Various studies proposed methods to inhibit the growth processes of iron and sulphate reducing bacteria which boost millions of times the reactions leading to iron toxicity of rice (Watanabe and Furusaka 1980, Ouattara and Jacq 1992, Dianou et al. 1998). Agronomy continue to seek for rice cultivars derived either from *Oryza sativa* – "the Asian rice" – or *Oryza glaberrima* – the "African rice" –, that can resist ferrous iron concentrations (Sahrawat 2004); pedology strives to propose various soil amendments which can help the crop to develop further resistance to intrusion into rice physiology . Subsurface drainage was proposed as solution to the problem in coastal floodplains and mangroves but led to situations of aggravation of acidification of pyrite soils, which is also a huge infertility factor (Breemen 1992, Suryadi 1996). Though some research had been performed in the field (Keïta et al. 2013b) it appears that the effect of subsurface drainage on hematite soils found in Tropical Savannah valley bottom irrigated rice soils – affected by iron toxicity – has not been investigated through experiment set according the universal principles of the scientific "design of experiments" (Figure 10.1). Such investigations were performed on ferrous iron intoxicated soils of Moussodougou, a valley bottom irrigated rice field located at the heart of the Tropical Savannah of Burkina Faso, West Africa.

The research was implemented using two parallel sets of experiments to measure the soil alteration and the intoxication level of the rice. These experiments were the buckets and the microplots experiments. These operations aimed to scientifically establish the relationship between on one side two factors – subsurface drainage and lime incorporation in the soil – and on the other side the soil behaviour measured through four responses: i) the ferrous iron concentration Fe^{2+}, ii) the soil acidity pH, iii) the oxidoreduction potential ORP, and iv) the dissolved oxygen DO. IRRI Iron toxicity scores (International Rice Research Institute (IRRI) 2002, Chérif et al. 2009) indicating the extent of the rice intoxication were also recorded during the crop growing period. The rice used was FKR19 (Institut de l'Environnement et de Recherches Agricoles (IN.ERA) 2000), a cultivar well know by farmers in Burkina Faso.

Figure 10.1: Flooded farm plot with submerging ferric iron in Moussodougou

Mr. Keita (right) is discussing with three rice production farmers of Moussodougou about iron toxicity. In front, resurging at the surface of ponding water in a farm plot, the ferrous iron precipitates as iron(III) hydroxides.

10.1. Material and methods

10.1.1. Bucket experiment design

The group of buckets used in this experiment was configured in order to allow subsurface drainage flow regulation, but also the soil acidity alteration. For this purpose, eight plastic buckets were gathered in the experimental site of Kamboinsé, each one having a volume of 15 litres. Four of the buckets (B1, B3, B5 and B7) received a tombac tap, fixed 5cm above their bottom (Figure 10.3). These taps were used to regulate the daily subsurface drainage flow through the ferrous iron intoxicated soil contained in the buckets, and brought from the valley bottom of Moussodougou. Afterwards, the eight buckets were separated into two sets. The first set included four buckets – B1, B2, B3 and B4 – for which two were drained at *Drng+* = 10mm/day and two were not drained, indicated by *Drng-* = 0 mm/day. *Lime* – $Ca(OH)_2$, known as a soil *pH* increase factor (Haynes and Naidu 1998) – was not incorporated in none the buckets of this first set (Figure 10.3). Similarly, the second set of buckets (B5, B6, B7, and B8) included two drained and two not drained ones. However, in contrary to the first set, in all these four buckets lime was incorporated in the soil.

The soil introduced in the eight buckets (the same was also used in the twelve microplots) was selected to be as close as possible to the contaminated soil found in the valley bottom irrigation scheme of Moussodougou. In spite of the long distance of 495 km, more than 25 m³ of soil were extracted in Moussodougou (Figure 10.2) and transported to the experiment site of Kamboinsé, located 15 Km in north-west of Ouagadougou, capital of Burkina Faso. Although the microbiological and the chemical composition of the soil was presumed to change during the transportation process – due to the exposition of the soil to air before submersion in the microplots – these characteristics were supposed not to alter to an extent that would prevent iron toxicity to take place in the experiments.

Figure 10.2: Extraction of the ferrous iron intoxicated soil at Moussodougou

The soils for the microplot experiments were extracted from 3 different locations in the valley bottom: upstream, middle and downstream. Farmers were fully involved the identification and helping for the extraction.

Table 10.1 Main characteristics of the soil of Moussodougou

Variable	Sample size	Mean			StDev	Variance	CoefVar(%)	Median	IQR	Skewness
%Clay	27.0	31.0	±	5.1	13.6	183.5	43.8	29.4	21.1	0.6
%Sand	27.0	49.4	±	6.3	16.8	282.5	34.0	54.7	21.6	-0.8
Db density	27.0	1.5	±	0.1	0.2	0.1	15.8	1.5	0.3	-0.3
%OM	27.0	7.0	±	1.6	4.2	17.6	60.3	5.8	4.7	1.2

Legend: StDev is the standard deviation and IRQ the interquartile range.

In order to reduce bias in the experiment results due to the variation of the soil properties, samples were extracted from several places in the valley. The soil of Moussodougou, with a mean of 27.0±5 % of clay and 49.4 ± 6.3% of sand, is a clayey soil according to USDA classification (United States Department of Agriculture (USDA) 1997). The soil presents some variations. Variations can be seen also in the values of the organic matter content %OM, though a mean of 7.0±1.6% (Table 10.1) is quite high for the hematite soils met in Tropical Savannah (Hillel et al. 2004). As one could expect, the dry bulk density is more or less stable to 1.5±0.1. Because of these inherent variations, the soil for the experiments was taken from three different locations in the valley: upstream, middle and downstream cross section Figure 10.2. The first 50 cm of top soils were collected.

10.1.2. Implementation of the bucket experiment

The initialization of the experiment process followed four principal steps. First, the soil in the eight buckets was flooded with a borehole irrigation water of pH 6.5 and hydrochloric acid was added to each one to take the pH to 3. Secondly, during five days, no operation was conducted in order to let the soil reach a new equilibrium in the buckets. A check was done to see and confirm that the pH remained at pH 3 from day-five to day-seven. Thirdly, at day-seven, lime was incorporated to the soil of four buckets (B5 to B8) at the rate of 1kg/m² of soil surface for each pH increment of +1 (Materechera et al. 2002, Caires et al. 2007). This way in these four buckets soils were brought up to pH 7. Finally, this very day, the experiment began by draining in random order only four buckets (B1, B3, B5 and B7, see Figure 10.3) and this process went on for 71 days. During these 71 days, the pH and the oxido-reduction potential (ORP) were measured once a day in a randomized order (Table 10.2).

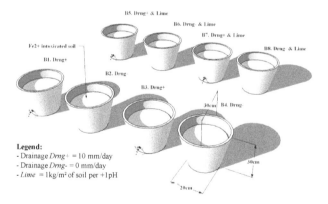

Figure 10.3: The Bucket experiment design

The figure shows the implementation replication principle in the 2²-factorial design. The front series of buckets did not receive any lime, while the backward series soil was mixed with lime. The couple of buckets (B1, B3), (B2, B4), (B5, B7) and (B6, B8) – having identical experimental conditions – are the replicates in which the soil pH and the redox potential ORP were measured during 71 non-consecutive days.

10.1.3. Statistical analysis of the bucket experiment

The fact that this bucket experiment is a 2²-factorial design (Mathews 2005) that takes place during a period of nearly 71 days exposed it the possible presence of a "lurking variable" associated with the time factor. The real two factors within which the experiment strove to compare the levels are the subsurface drainage conditions – with two levels: *Drng-* = 0 mm/day, and *Drng+* = 10 mm/day – and the liming conditions – with also two level Lim- = 0 lime incorporated, *Lime+* = lime incorporated at the rate of 1 kg/m² of soil surface for each *ΔpH*=+1. However, since the flooded rice soil evolves also in time, this time dimension brings a third factor with 71 days levels, and more adversely, the impossibility to select these factor levels randomly. In order to solve this difficulty of the lurking variable impacting the soil *pH* and *ORP* along the timeline, the time factor was blocked. Hence, the experiment was a so the called *randomized block design*. The main consequence is that not statistically legitimate claim (Montgomery 2001, Antony 2003) can be done about the effect of the time on the soil response measured through the *pH* and the *ORP*. However, the ANalysis Of VAriance model (ANOVA) would provide valid inferences about the impacts of the drainage levels and the lime incorporation on the two responses which are the *pH* and the *ORP*.

Table 10.2 Bucket experiment design matrix

Std Order	Run Order	Day	Drainage	Lime	Replicates	Legend
1	3	1	Drng-	Lime+	2	Dates
2	2	1	Drng+	Lime+	2	1 = 08 Jan 2014;
3	4	1	Drng+	Lime-	2	10 = 17 Jan 2014; etc.
4	1	1	Drng-	Lime-	2	Drainage & Depth
1	3	10	Drng-	Lime+	2	Drng- = 0 mm/day
2	4	10	Drng+	Lime+	2	Drng+ = 10 mm/day
3	1	10	Drng+	Lime-	2	Lime - = 0 Kg/m²
4	2	10	Drng-	Lime-	2	Lime + = 1 kg/m² per increment +1 pH

The table shows the 2² factorial design matrix for two dates (Jan 8th and Jan 17th) out of a total of 35 different dates (71 days). The standard order is also shown. The implemented run order complied with the randomisation requirement of experiments, so that statistically valid claims could be made as for the impact of the drainage two levels and the lime two levels on the responses measured (*pH* and *ORP*).

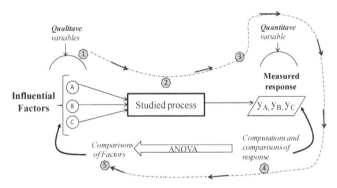

Figure 10.4: ANOVA model principle

The comparison between qualitative variables (the factors) is done through the use of a quantitative variable (the response). ANOVA is a statistical model based on this principle. It allows the comparison of a process response influential factor (A, B, C) through comparisons of the measured response (*y*). The numbers 1-5 indicate the different steps of the research performed about the studied process. ANOVA is applied at the two steps 4 and 5 and will allow comparing the factors.

Although the statistical software Minitab v.16 (Mathews 2005) was used to analyse both the bucket and the microplot experiments, a correct computation and interpretation of the results require the description of the principles and the formula used in the ANOVA model. The basic problem ANOVA aims to solve is that it is not possible to directly compare qualitative variables or factors. The comparison, to be statistically valid, is done through the use of a quantitative variable – the *response* in the experiment – associated with the factors (Sheskin 2004). Variations around a process response means (Figure 10.4) are quantified by the variances. To compare two or more means, ANOVA creates the *F-statistic* based on the variances (Mathews 2005). Subsequently, it compares the computed *F* with critical levels to make inferences about the significance of the differences between the response means. If these differences are significant, then valid claims can be made about the differences between the (qualitative) factors.

The *F-statistic*, the corner stone of ANOVA, is made of a ratio of two variances: one related to the treatment means, and the second related to the population means. According to the *central limit theorem* (Montgomery 2001, Antony 2003), if all the treatments were equal – i.e. if all the treatments were drawn from a unique population –, the *F* ratio would be equal to 1 (or a critical value). Basically, this ratio is expressed as follows:

$$F = \frac{Between\ variance\ of\ treatments}{Within\ variance\ of\ treatments} \qquad (10.1)$$

where: variance is expressed in term of *sum of squares*; a *treatment* is made of all the measured response values associated with a specific factor level, therefore also including the replicates.

In a design using only one factor "*A*" with k levels $a_1, a_2,.., a_k$, the *F-statistic* calculation follow *four* main steps (Figure 10.5). Since each of the factor level is equivalent to a specific experimental condition, each factor level is also called a *treatment*. Therefore, $a_1, a_2, ..., a_k$ form k distinct treatments. If each treatment is identically reproduced n times (in order to reduce the measurement errors), the number n is the number of *replicates*. Hence, a treatment is made of n replicates. (Mason et al. 2003). When a treatment is set in the experiment – i.e. a factor level is initialized – and the response y values are measured, three kinds of deviation of the response from the mean – called *sum of squares* – can be defined (Montgomery 2001).

1) The sum of squares (over the k treatments) of the gap between all the individual means of treatment $\overline{y}_i$ and the grand mean $\overline{\overline{y}}$ is first calculated. Subsequently, this value is multiplied by the number of replicates n to yield the *treatments sum of squares*:

$$SS_{treatments} = n\sum_{i=1}^{k} (\overline{y}_i - \overline{\overline{y}})^2 \qquad (10.2)$$

One can notice that Eq.(10.2) expresses the variation *between* the k distinct treatments and then adds them up. Therefore this expression is sometimes noted $SS_{between}$ (Figure 10.5, (a)).

2) The sum of squares of the gap between all the n individual y_{ij} responses and the mean of the responses within treatment i is first calculated. Subsequently, these values are added for all the k treatments to form the *error sum of square*:

$$SS_{error} = \sum_{i=1}^{k}\sum_{j=1}^{n}(y_{ij} - \bar{y}_i)^2 \qquad (10.3)$$

One can notice that Eq.(10.3) expresses the variation *within* the k distinct treatments and then adds tem up. Therefore this expression is sometimes noted SS_{within} (Figure 10.5, (b)).

3) The sum of squares (over the k treatments and the n replicates) of the gap between all the individual responses y_{ij} and the grand mean $\bar{\bar{y}}_i$ is first calculated. This yields the *total sum of squares*:

$$SS_{total} = \sum_{i=1}^{k}\sum_{j=1}^{n}(y_{ij} - \bar{\bar{y}}_i)^2 \qquad (10.4)$$

It can be shown (Balley 2008) that: $SS_{total} = SS_{error} + SS_{treatments}$.

The four steps of the *F-statistic* calculation are as follows. In the first step the equations Eq(10.3), Eq.(10.2), and Eq.(10.4) are computed (see also Figure 10.5). The *second step* consists of the calculation of the *degrees of freedom (df)* of each of the sums of squares. Typically, each statistic computed for a data set "consumes" one degree of freedom from the original data set of size N. For example, the variance has a degree of freedom of N-1 since one unit is "consumed" by the computation of the mean prior to the calculation of this variance (Mathews 2005). At the *third step*, the variances associated with each of the three sums of squares are determined as the ratio of the sum of squares and the degree of freedoms. These variances are called *mean squares (MS)*. For a one-factor A design with k treatments each one having n replicates, the two related mean squares are given by:

$$MS_{treatments} = \frac{SS_{treatments}}{df_{treatments}} = \frac{SS_{treatments}}{k-1} \qquad (10.5)$$

$$MS_{error} = \frac{SS_{error}}{df_{error}} = \frac{SS_{error}}{k(n-1)} \qquad (10.6)$$

Finally in agreement with the meaning of Eq.(10.1), the *F-satistic* is expressed in this case as the ratio:

$$F_A = \frac{MS_{treatments}}{MS_{error}} \qquad (10.7)$$

In a multiple factors experiment several additional terms are calculated by the software. For example, if the two factors are A and B (like in the bucket experiment), F_A and F_B are calculated and will indicate – through the comparison with the critical values of F for the related degree of freedom – whether the main effect of the factors are significant or not. A *p-value* is associated with these comparisons. In addition, the interaction between factor A and B to impact the response may be calculated. These *F-statistics* are summarized by the software MINITAB in the so called ANOVA Table (see Table 10.3).

Table 10.3 Two-way ANOVA table

Source of variation	Sum of squares	Degree of freedom	Mean square	F-statistic
A	SS_A	df_A	MS_A	F_A
B	SS_B	df_B	MS_B	F_B
AB	SS_{AB}	df_{AB}	MS_{AB}	F_{AB}
Error	SS_ε	df_ε	MS_ε	
Total	SS_{total}	df_{total}		

Source: adapted from Mathews (2005)

Three more operations done with the software Minitab to analyze the microplot experiment data need to be mentioned. First, before the application of the ANOVA model, the three required conditions that prove its validity – the normality, homoscedasticity (equality of the variances) and the independence of the response variable (*pH* and *ORP*) –were checked. However, on should notice that slight deviations from these conditions do not compromise de the validity of the results since the ANOVA is not very sensitive to departures from normality and homoscedasticity (Boslaugh and Watters 2008, Rumsey 2009). Secondly, it must be also noticed that most often, the *pH* distribution from rough data is not a normal distribution. The *pH* distribution has been an object of a remarkable controversy among researchers (Boutilier and Shelton 1980). To recover normality of the data to some extent, and allow a legitimate application of ANOVA, the following transformation was performed on the *pH* data:

$$\text{Transf}(\text{Log}_{10} pH) = \text{Log}_{10} pH \text{-MEDIAN}(\text{Log}_{10} pH) \qquad (10.8)$$

The reverse function to get the *pH* from these values given by:

$$pH = 10^{\left[\text{Tranf}(\text{Log}_{10} pH)+\text{MEDIAN}(\text{Log}_{10} pH)\right]} \qquad (10.9)$$

where: *Transf(Log₁₀ pH)* represents the transformed response.

Finally, after the ANOVA computation, the so-called ANOVA post test is performed to check the significance of the differences between pair of treatments. This was indeed the main objective of the analysis since it would tell whether a specific treatment created a significant effect on the measured response or not. These tests were performed using Tukey method (Tukey 1977, Hoaglin et al. 1991).

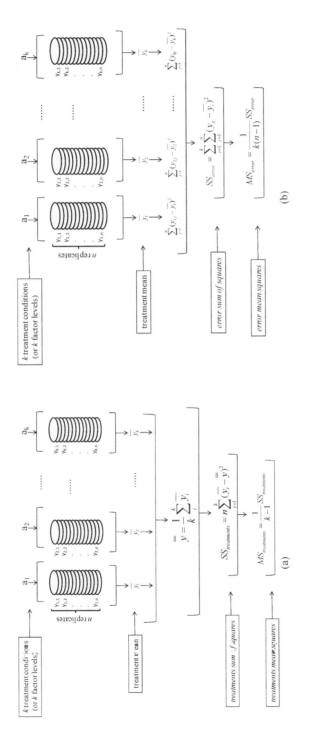

Figure 10.5: The computation procedure of the F-statistic, corner stone of ANOVA

Experiment or survey design with a unique factor A having k treatment conditions a_i, each of which has n replicates. One can see that the *treatments mean squares* evaluates the variance *between* the treatments (a). On the other hand, the *error mean squares* evaluates the variance *within* the treatments (b). The *F-Statistic* is the ratio of variances $F = MS_{treatments}/MS_{error}$ which, if all the treatment samples were coming from the *same population*, would be equal to *one* according to the central limit theorem. If the F is greater than *one* (or a critical value), then there is at least one treatment sample coming from a population different from the others. This procedure was used with the current bucket and the microplot experiments as well.

10.1.4. Microplot experiment design

The microplots were constructed to resist temperature changes and to offer a space convenient for rice full growing. The material used is reinforced concrete to prevent cracks due to temperature variation – from 15°C to 45 °C – following the alternation of the rainy and the dry seasons In Burkina Faso. A special chemical substance called *Sikalite* (Bel Ombre Development Company (BODCOLTD) 2011) was incorporated in the mortar in order to ensure the impermeability of the walls. On the other hand, based on the average rice rooting depth of 50 cm, the dimensions were chosen as follows (Figure 10.6a):

- the height H= 1.50 m; the top and the bottom are square of side Sd =1.00 m;
- the wall thickness amounts to e = 8 cm; the bottom thickness is b = 15 cm;
- and the freeboard representing the leftover height after the microplot is full of soil and ponding water F = 5 cm.

Therefore, the total available internal volume was set to 1.30 m^3, offering enough space for 4x4 rice tillers transplanting at a spacing of 20 cm.

The other components of the microplots were made of tombac, PVC and iron contaminated soil. For these purposes, one of the walls of the microplot possesses two taps. The first tap was connected to 40 mm-φ PVC pipe on the top-generator line of which are perforated small orifices of 5 mm-φ with a 15 mm spacing (Figure 10.6-(b)). This PVC pipe is given a zigzag shape and is supported by seven 30 cm-high small pillars rising above the microplot's internal bottom. The 30 cm space between the drainage pipe and the microplot internal bottom is filled with sand that rises some 5 cm above the pipe and is aimed to facilitate the water flow into the perforations. One end of this drainage pipe crosses the wall and is ended with a flow adjustable tombac tap outside the concrete wall. A second tap, to which not PVC pipe is connected inside the microplot, is mounted though the same wall at 85 cm above the first one. It was used to drain the ponding water layer under the rice (Figure 10.6). The soil introduced in the microplots is the same soil of Moussodougou used in the bucket experiment.

10.1.5. Implementation of the microplot experiment

The rice used in the nursery was selected for his appreciated characteristics and cycle length by the farmers. The cultivar is locally known as FKR19 and was developed by INERA of Burkina Faso from the Asian rice specie *Oryza sativa*. This cultivar has a 85 days growing cycle and was used by more than 90% of the farmers in the valley bottoms irrigated rice schemes of Tiefora and Moussodougou (Sokona et al. 2010). The main reasons behind this success are its capability to adapt both to rainy and dry seasons, the good quality of the husked grain – grain length: 9.3 mm, width: 2.5 mm, 1000-grain weigh: 25.3g –, its availability, and its average yield of 5-7 tons per ha (Institut de l'Environnement et de Recherches Agricoles (IN.ERA) 2000). To avoid quality degeneration and yield drops, farmers renew the seed every two growing seasons. The recommended nursery period of 3 weeks was adopted for the microplot experiment.

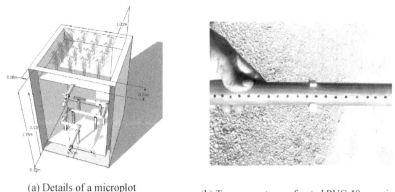

(a) Details of a microplot

(b) Top-generator perforated PVC 40mm pipe
used for subsurface drainage

Figure 10.6: Components of the microplot used in the experiment

In part (a), the cross section of a reinforced concrete microplot shows the space from the bottom
to the top (normally full with ferrous iron contaminated soil), the shape and location of the bottom
drainage PVC pipe, and the two adjustable drainage taps. In part (b), the 5mm-ϕ orifices are
shown on the drainage pipe.

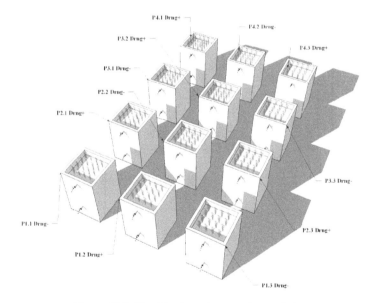

Figure 10.7: The 12-microplot experiment design

Legend: *Drng+* = 10 mm/day; *Drng-* = 0mm/day. The notation *"Pi.j"* identifies the microplots, *i*
representing the line, and *j*, the column. The microplots with *Drng+* = 10 mm/day were
permamently drained during the whole experimental period of 86 days. The responses (Fe^{2+}, *pH,
ORP, and DO*) measurements were performed in a randomized order on the microplots at
Depth- = 30 cm and *Depth+* = 50 cm.

Table 10.4 Microplot experiment design matrix

Std Order	Run Order	Day	Drainage	Depth	Replicates	Legend
1	3	1	Drng-	Depth-	18	Dates
2	1	1	Drng-	Depth+	18	1 = 05 Feb 2014; 23 = 28 Feb 2014
3	4	1	Drng+	Depth-	18	58 = 04 Apr 2014; 86 = 02 May 2014
4	2	1	Drng+	Depth+	18	Drainage & Depth
5	6	23	Drng-	Depth-	18	Drng- = 0 mm/day
6	8	23	Drng-	Depth+	18	Drng+ = 10 mm/day
7	7	23	Drng+	Depth-	18	Depth- = 30 cm
8	5	23	Drng+	Depth+	18	Depth+ = 50 cm
9	12	58	Drng-	Depth-	18	Day = number of days from rice transplantation
10	9	58	Drng-	Depth+	18	Replicates
11	10	58	Drng+	Depth-	18	Total of Microplots : 12
12	11	58	Drng+	Depth+	18	- Six are Drng- (drainage = 0 mm/d)
13	14	86	Drng-	Depth-	18	- Six are Drng+ (drainage = 10 mm/d)
14	16	86	Drng-	Depth+	18	Measurements of the response in 3 random soil extracts for
15	15	86	Drrg+	Depth-	18	each of the 6 microplots (Drng- and Drng+) at depth- and Depth+,
16	13	86	Drrg+	Depth+	18	in random order. Therefore, there are 3*6 = 18 replicates for each run

Alongside the nursery growing, twelve microplots were prepared and the measurements performed in the strict respect of the randomness and replication principles. The 12 microplots were organized in four lines and three columns and labeled per line from P1.1 to P4.3, the first index indicating the line and the second, the column (Figure 10.7). Hence, starting from P1.1, a microplot was marked for drainage ($Drng+ = 10 \ mm/day$) and the following for non-drainage ($Drng+ = .0 \ mm/day$), alternating this way up to P4.3. After the completion of four week, the rice was transplanted into the microplots with a spacing of 20 cm x 20 cm. The initial measurements of the four variables – ferrous iron concentration Fe^{3+}, soil acidity pH, oxido-reduction potential ORP, dissolved oxygen DO, and IRRI iron toxicity scores (International Rice Research Institute (IRRI) 2002) – being investigated were carried out, proceeding randomly (Table 10.4) with the two depths ($Depth- = 30 \ cm \ and \ Depth+ = 50 \ cm$). Subsequently, these measurements were performed mainly at three other non consecutive dates, respectively: 23, 58, and 86 days after rice transplantation in the microplots.

Various procedures were used to measure the response variables. For the ferrous iron concentrations (the most important response in this experiment), soil samples were extracted at the two depths of 30 cm and 50 cm (Keïta 2014), and a reflectometer with Fe^{2+} reflectoquant strips test was applied (Keïta 2013b). A pH-meter/oximeter Hach HQ40D – on which were mounted a pH/ORP probe Intellical PH101 and a dissolved oxygen probe LDO101 – was used to measure the soil acidity pH, the oxido reduction potential ORP and the dissolved oxygen DO (Bier and Lange 2009). For iron toxicity scoring, the method, based on the observation of the proportions of bronzing leaved at different rice growing stages was applied (International Rice Research Institute (IRRI) 2002).

10.1.6. Statistical analysis of the microplot experiment

Except that the responses being measured and the frequency of measurements, the statistical analysis applied in the microplots experiment is widely similar the one used in the buckets experiment. Basically, the ANOVA General Linear Model was performed using the statistical software package Minitab. The design is a 2^2-factorial, but unbalanced design due to the withdrawal of some outliers presenting extreme low or high values. The two factors impacting on the response Fe^{2+}, were the subsurface drainage and the depth in the rootzone. These two factors were analysed by ANOVA, while the three other responses – DO, pH, ORP –, and iron toxicity scores were described only in terms of their evolution in time by Boxplot, due to the non-normality of the data. The refined analyses of the pH and ORP were already performed in the buckets experiment, in a much more confined medium.

10.2. Results and discussion

10.2.1. Pre-ANOVA requirement checks for both experiments

After the application of some transformations on the rough data, the ANOVA three precondition checks were satisfactorily met. These transformations were, for the pH, the one described in Eq.(10.8), and for the ORP, the logarithm base ten. The operations yielded symmetrical Boxplots (Tukey 1977), though slightly bigger for the pH at the treatment condition $Drng$-/$Lime$- (Figure 10.8 and Figure 10.9). This higher size is often met in unbalanced experiments due to some missing values that led to unequal sample sizes per treatment condition. It does not imply heteroscedasticity, and therefore, all the transformed *pH* and *ORP* data were considered as having the same variance for all the related treatment conditions (Montgomery and Runger 2011). On the other hand, the symmetry, conveying the normality of the transformed data, was confirmed by the linear normality plots and the Anderson Darling (A-D) test p-values all smaller than 5% (Anderson 1941, Boslaugh and Watters 2008). The independence of the data was also successfully checked separately for the four treatments conditions for *pH* and *ORP*. However, it should be noticed that the time variable is not shown, since the incapability to randomize the time makes it difficult to pretend valid claims by ANOVA about their impact on the measured responses *pH* and *ORP* (Mathews 2005). Hence, the time variable was "blocked" during the analysis (Balley 2008).

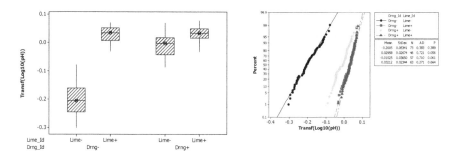

Figure 10.8: Boxplot and normality check of Transf(Log$_{10}$(pH)) in the buckets experiment

The Boxplot with symmetric medians and comparable sizes, and the normality plots indicate the normality and the equity of the variances of the data in the various treatments conditions, and allow the use of ANOVA.

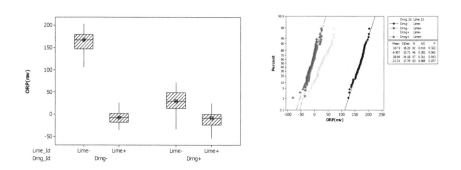

Figure 10.9: The normality check of reduction potential ORP in bucket experiment

Normality and homoscedasticity (equality of variances) are visible in the two plots, ensuring the legitimacy of ANOVA use to treat data.

Similarly, the response variables were checked convenient for appropriate the ANOVA computations with the microplots experiment. As it can be seen on Figure

10.10, the box size are equivalent, except around date 58, where they are bigger due to higher sample sizes for the related treatment conditions. However, as pointed out by Mathews (2005), the differences in box sizes due to unbalanced data (different sample size) does not illegitimate the use of ANOVA as if the samples were of equal variances. The normality linear plot and the Anderson-Darling p-values- all smaller than 5% - confirmed that the data follows a normal distribution (Rumsey 2009). On the other hand, the scatterplot of $Log_{10}(Fe^{2+})$ does not display any particular pattern, supporting the independency of the measurements, and completing ANOVA requirement checks.

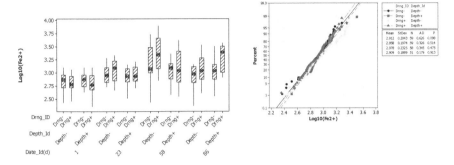

Figure 10.10: Boxplot and normality check of $Log_{10}(Fe^{2+})$ in the microplots experiment

The symmetry is visible for most of the boxes and the normality of the transformed data is further confirmed on the normality linear plot. The presence of three boxes of longer size at date 58 days are due to greater sample sizes (unbalanced experiment) for the related treatments and do not illegitimate the use of ANOVA as if the samples are of equal variance (Montgomery 2001, Mathews 2005).

10.2.2. Post-ANOVA requirement checks for both experiments

After running the ANOVA, the post-ANOVA tests for its legitimacy were well confirmed by the plots of the residuals, i.e. the gaps between measured responses and the model predictions. These plots are required to pretend to any legal inference about the results in a comparative experiment (Mathews 2005, Balley 2008). The normality, the constancy of the variances and the independency of the residuals are well supported for $Log_{10}(Fe^{2+})$ in Figure 10.11. On the "residual vs. fitted value" (Figure 10.11-(C)), the columns of dots correspond to the response $Log_{10}(Fe^{2+})$ related to the four blocked dates (1, 23, 58, and 86). However, while the constancy of variances (Figure 10.12-(C)) and the independency (Figure 10.12-(D)) are insured, there appears a slight deviation of the residuals from normality with the transformed *pH* response residuals (Figure 10.12-(A) and (B)). As already stated, this does not question the validity of the ANOVA since it is not sensitive to slight deviation from normality (Montgomery 2001, Rumsey 2009). Similarly, the post ANOVA tests were also successful with transformed *ORP* residuals as it can be observed on Figure 10.13. The model outputs are therefore valid.

10.2.3. Increase in ferrous iron but decrease in soil acidity

The results of the microplot experiment show that subsurface drainage definitely alter the ferrous iron concentration in the soil, independently from the depth and the timeline. The *F-statistic* yielded 5.87, with a high significance level of $p = 1.6\%$, and the Tukey test confirmed that the impacts of the two levels of the drainage factor are significantly difficient (Table 10.5 and Table 10.6). The same conclusion is displayed by the main effect plots (Figure 10.14). Concretely, a drainage rate of 10 mm/day led to the increase of ferrous iron concentration from 935 mg/l to more than 1105 mg/l at 95% of confidence level. (Table 10.6). The variation of the ferrous iron concentration within the depth range of 30 cm – 50 cm is not significant, as it is shown by the important p value of 33.4% and the main effect plot. Furthermore, there is no significant interaction between the subsurface drainage and the depth factors as for impacting on the ferrous iron concentration. The corresponding p value is 45.2%, much higher than 5.0%. There was also a sense that – though this is not a legitimate claim because of the impossibility to randomise –the ferrous iron concentration had increased as time went on during the rice growing season. It is clear that subsurface drainage increased the level of ferrous iron in the soil in both depths 30 cm and 50 cm.

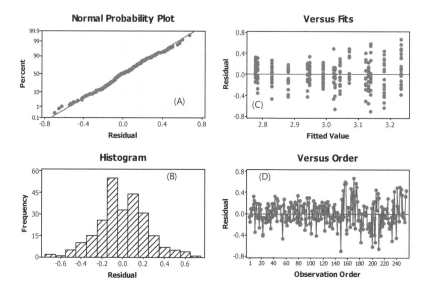

Figure 10.11: Residual Plots of $Log_{10}(Fe^{2+})$ for ANOVA Check.

Microplots experiment. The normality (A, B), the constancy of variance (C),
and the independency of the residuals (D) are confirmed by the four plots.
This supports the legitimacy of the use of ANOVA model.

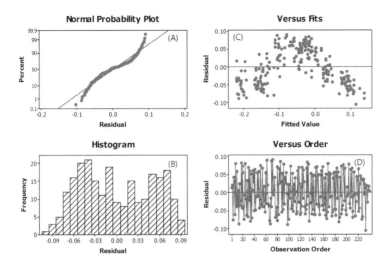

Figure 10.12: Residual Plots of Transf(Log₁₀(pH)) for ANOVA Check

Buckets experiment.Only slight deviation from normality at the extremes (A, B), the constancy of variance (C), and the independency of the residuals (D) are confirmed by the four plots. This supports the legitimacy of the use of ANOVA model.

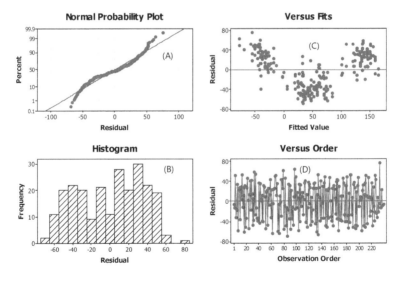

Figure 10.13: Residual plot of Log₁₀(ORP) for ANOVA General Linear Model bestFit check

Buckets experiment. Only a slight deviation from normality at the extremes (A, B), the constancy of variance (C), and the independency of the residuals (D) are confirmed by the four plots. This supports the legitimacy of the use of ANOVA model.

Table 10.5. Two-way ANOVA Table with Interaction for $Log_{10}Fe^{2+}$ using Adjusted SS for tests

Source of variation	Degree of freedom	Seq SS	Adjusted sum of squares	Adjusted Mean Squares	F	p(%)
Drng_ID	1	0.38	0.34	0.34	5.87	1.6%
Depth_Id	1	0.05	0.05	0.05	0.94	33.4%
Blck-Date_Id	3	4.47	4.48	1.49	26.16	0.0%
Drng_ID*Depth_Id	1	0.03	0.03	0.03	0.57	45.2%
Error	247	14.10	14.10	0.06		
Total	253.00	19.04				
S = 0.24		R-Sq = 25.92%		R-Sq(adj) = 24.12%		

Table 10.6: Grouping Information Using Tukey Method and 95.0% Confidence

Factor Level	Sample size N	Mean(Log₁₀Fe2+)	Fe^{2+} (mg/l)	Grouping[a]
Drng+	129	3.044	1106.62	A
Drng-	125	2.971	935.41	B

[a]Means that do not share a letter are significantly different.

Though these impacts of subsurface drainage by increasing ferrous iron concentration are rather surprising, they can be logically explained. In fact, it was expected that subsurface drainage will reduce ferrous iron concentration in the soil by bringing more oxygen in the root zone and incommode the expansion of sulphate and iron reducing bacteria (SRB and IRB), which are essentially – but not strictly – anaerobic bacteria (Ottow and Glathe 1971, Ouattara 1992). However, as seen with the field measurements performed on the soil of Moussodougou (see Section 6.7) and shown by Keïta et al. (2013b), the Tropical Savannah valley bottom soils are mainly formed by hematite – and not pyrite as often found in coastal floodplains and mangroves (Suryadi 1996). Therefore, the key geochemical reactions involved are oxidation and the reduction of hematite, in association with specific bacteria. As already mentioned in Section 6.7, the two reactions are the following:

- Formation of hematite (oxidation): Decrease of Fe^{2+} and increase of acidity (Schwertmann U. and Murad E. 1983, Kato et al. 2008):

$$4Fe^{2+} + O_2 + 4H_2O \rightarrow 2Fe_2O_3 + 8H^+ \qquad (10.10)$$

- Reduction of Hematite: Increase of Fe^{2+} and decrease of acidity (Breemen 1992):

$$2Fe_2O_3 + CH_2O + 8H^+ \rightarrow 4Fe^{2+} + CO_2 + 5H_2O \qquad (10.11)$$

A first way to clearly determine which of the two above equations is involved in the experiments related to the soil of Moussodougou, is to confront them with the concomitant variations of pH and ORP with Fe^{2+} as revealed by the measurements. While Eq.(10.10) consumes ferrous iron with an increase of acidity, Eq. (10.11) increases ferrous iron, with a simultaneous decrease of acidity. In fact both the bucket and the microplot experiments – conducted independently – led to an increase of the pH of the soil (Figure 10.16 and Figure 10.17). These simultaneous increase of Fe^{2+} and pH comply with Eq. (10.11): there was reduction of the hematite, with – as suggested by Keïta et al. (2013b) – strong involvement of iron reducing bacteria (IRB). The reduction is also attested by the decrease – from 84.6 to 9.2 mV, at 95% confidence level – of oxidoreduction potential ORP (Table 10.10 and Figure 10.15), and the decrease of dissolved oxygen DO from more than 1 mg/l to less than 0.1 mg/l (Figure 10.18) all along the rice growing period.

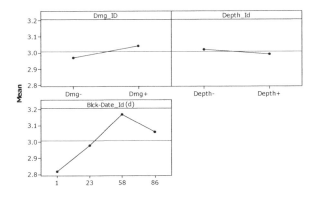

Figure 10.14: Main effect plots of the factors of $Log_{10}(Fe^{2+})$

Subsurface drainage increases the ferrous iron concentration in the hematite submerged soil. On the other hand, there is no significant difference between the two depths 30 cm and 50 cm in Fe^{2+} concentrations. The increase of Fe^{2+} from along the crop growing timeline is also shown with the *Blck-Date_Id* expressed in days, though this factor was not randomised.

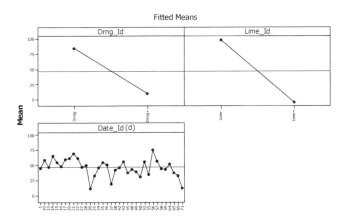

Figure 10.15: Main effect plots of the factors of $Log_{10}(ORP)$

Subsurface drainage decreased the oxidoreduction potential (ORP) from 84.6 to 9.2 as differential effects (related to non-drained and drained conditions) in the soil. So did the lime incorporation. The fluctuation with no clear trend of ORP as effect of the time factor is also displayed.

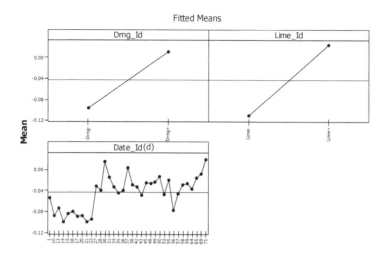

Figure 10.16: Main effect plots of the factors of Transf(Log₁₀(pH)) in bucket experiment

Subsurface drainage increases the *pH* from 5.7 to 7.3 (using the inverse transformation
$10^{\wedge}(\text{Transf}(\text{Log}_{10}pH)+\text{Median}[\text{Log}_{10}pH])$, so do the lime incorporation into the soil. The *pH*
increase along the timeline is also visible.

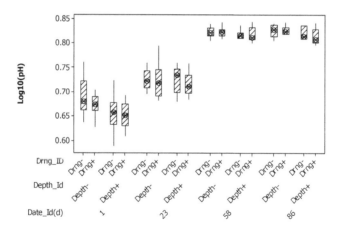

Figure 10.17: Boxplot of Log₁₀(pH) in the microplot experiment

The increase of *pH* (reduction of acidity) is visible in the hematite soil along the crop growing
season, though this figure does not show the differential effects in drained and non-drained
microplots.

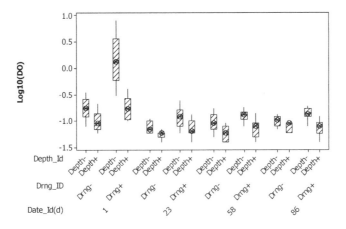

Figure 10.18: Boxplot of Log₁₀(DO) in the microplot experiment

The decrease of *DO* (dissolved oxygen) is visible in the hematite soil along the crop growing season from more than 1 mg/l to less than 0.1 mg/l, though this figure does not show the differential effects in drained and non-drained microplots.

Table 10.7: ANOVA Table with for *ORP*, using Adjusted SS for tests

Source of variation	Degree of freedom	Seq SS	Adjusted sum of squares	Adjusted Mean Squares	F	p (%)
Date_Id	34	57660	47345	1390	0.96	54.3%
Drng_Id	1	495640	47345	319700	219.51	0.0%
Lime_Id	1	573235	47345	573235	393.58	0.0%
Error	202	294200	47345	1456		
Total	238	1420730				
	S = 38.16		R-Sq = 79.29%		R-Sq(adj) 0.756	

Table 10.8: Two-way ANOVA Table with for *Transf(Log₁₀(pH))*, using Adjusted SS for Tests

Source of variation	Degree of freedom	Seq SS	Adjusted sum of squares	Adjusted Mean Squares	F	p (%)
Date_Id	34	0.25	0.26	0.01	2.56	0.0%
Drng_Id	1	0.97	0.65	0.65	218.04	0.0%
Lime_Id	1	0.98	0.98	0.98	331.15	0.0%
Error	202	0.60	0.60	0.00		
Total	238	2.80				
	S = 0.05		R-Sq = 78.63%		R-Sq(adj) = 74.82%	

Table 10.9: Grouping Information Using Tukey Method and 95.0% Confidence

Factor Level	Sample size N	Mean(Transf(Log₁₀pH))	pH	Grouping[a]
Drng+	120	0.01	7.3	A
Drng-	119	-0.10	5.7	B
Lime+	109	0.02	7.6	A
Lime-	130	-0.11	5.6	B

[a]Means that do not share a letter are significantly different.

Table 10.10: Grouping Information Using Tukey Method and 95.0% Confidence

Factor Level	Sample size N	Mean(ORP) in mV	Grouping[a]
Drng-	119	84.57	A
Drng+	120	9.25	B
Lime-	130	98.35	A
Lime+	109	-4.53	B

[a]Means that do not share a letter are significantly different.

A second way to explain why Eq. (10.11) represents the right model of the chemical process in the hematite soil is to consider what shows the Nernst Equation (Murray 1994). In the current case, there is a decrease of the oxidoreduction potential *ORP* while the *pH* increases. The Nernst Equation – building the relationship between *ORP* and the *pH* – provides an explanation of this phenomenon. Given an oxidoreduction half cell reaction such as:

$$aA + bB + n[e^-] + h[H^+] = cC + dD \tag{10.12}$$

The Nernst Equation is stipulated as follow (Housecroft and Constable 2006):

$$ORP = ORP_0 + \frac{0.05916}{n} Log_{10} \left[\frac{\{A\}^a \{B\}^b}{\{C\}^c \{D\}^d} \right] - \frac{0.05916h}{n} pH \tag{10.13}$$

where: *a, b, n, h, c,* d are the number of moles of the chemical species *A, B, e⁻, H⁺ ,C* and *D*, respectively. *ORP* is the oxido reduction potential; *ORP₀* is the standard electrode potential ; the terms within round brackets represent the chemical activity of the species.

The Nernst Equation is a straight line with a slope of $-\dfrac{0.05916h}{n} pH$. When the proton *H⁺* is on the left side of the equation – such as in Eq.(10.11) or Eq.(10.12) – the *h* is positive, and then the slope is negative. In that case, high values of the *pH* will correspond to small values of *ORP*, and the ongoing process is a reduction process. This negative correlation between the *pH* and *ORP* is further confirmed in the microplot experiment by the scatterplot in Figure 10.19.

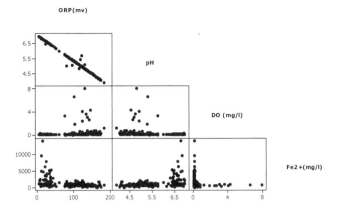

Figure 10.19: The matrix scatterplot of three potential quantitative factors of the response Fe^{2+} in the microplots experiment

From top to bottom, one can see the confirmation of Nerst's Equation (Eq.(10.13)) relating the oxidoreduction potential *ORP* and the hydrogen potential *pH*. In the present case, the negative slope denotes an ongoing reduction process in the soil of the experimental microplots. The plot also shows the surjective relationship between *pH* or *ORP* and *DO*, making the two first bad predictors of the last. One can also notice the more or less surjective relationship between *ORP* or *pH* and Fe^{2+} , which does not indicate them as good predictors of Fe^{2+} concentrations in the soil; and the relatively neat L-pattern relating the *DO* and Fe^{2+} concentrations, namely Fe^{2+} concentration increases when dissolve oxygen *DO* concentration decreases. These observations are confirmed at higher plotting scale of Fe^{2+} .

10.2.4. Improvement of soil resistance to iron intoxication

Despite of the fact that the more oxygenated irrigation water renewal after regular subsurface drainage did not prevent oxygen depletion is not in contradiction with the soil improvement. Sulphate and iron reducing bacteria (SRB and IRB) are not strict anaerobic (Traore et al. 1981, Dianou et al. 1998). It sounds logical to say that these bacteria simultaneously used the ferric iron provided by hematite and the oxygen brought by irrigation water as electron donors for their metabolism (Traore et al. 1982). As a result, ferric iron Fe^{3+} was transformed into ferrous iron Fe^{2+} ,(see Eq.(10.11)) while the oxygen content of the submerged soil was being depleted (Figure 10.18). But fortunately the soil acidy also simultaneously decreased as an effect of subsurface drainage. Even though in the current experiment the rice plants eventually all died due to a too high iron toxicity score of 6 to 7(Figure 10.20), the root intake of ferrous iron when the *pH* increases will gradually decrease (Breemen 1992). This process announces a progressive improvement of the soil as regards with ion toxicity.

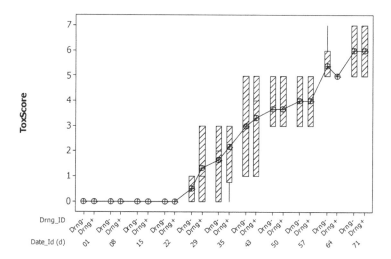

Figure 10.20: IRRI iron toxicity score has increased along the crop growing period

The increase of IRRI Fe^{2+} toxicity score (*ToxScore*) (International Rice Research Institute
(IRRI) 2002) is visible in the hematite soil along the crop growing season, though this figure
does not show the differential effects in drained and non-drained microplots. This increase
eventually led to the death of almost all the rice plants.

These experiments re-placed the *pH* – and not oxygen – as the central
characteristic to alter in order to solve iron toxicity issue in hematite soils. Even with a
drainage rate as important as 10mm/day, the oxygenation in the rootzone was not
enough to prevent the growth of iron reducing bacteria and the ferrous iron
concentration. However, it did help in increasing the soil *pH* from 5.7 to 7.3 ($p = 5\%$)
which will limit the absorption of ferrous iron by the rice and thus, improve rice
yield(Dave 1985). However, it must be mentioned that the same acidity reduction was
also achieved through the incorporation of lime in the soil at a rate of 1 kg/m², as shown
by the results of the bucket experiment (Figure 10.15; Table 10.9 and Table 10.10). The
pH increased by 2 units (from 5.6 to 7.6 at a significance level of $p = 5\%$). The impact
of lime as a soil acidity reducer is well known and has been widely used in acidic soils
(de Oliveira EL and Pavan 1996, Materechera and Mkhabela 2002). Lime (Ca(OH)$_2$),
should be preferred to gypsum (CaSO$_4$·2H$_2$O) for the *pH* improvement due to the
possible promotion of sulphate reducing bacteria BSR (Dianou 2005). The limited
availability of this mineral may make it less competitive against subsurface drainage.

10.3. Conclusions

Our results show that in hematite valley bottom Tropical Savannah soils subsurface
drainage i) increases ferrous iron concentration (935 to 1106 mg/l) and short term iron
toxicity, ii) decreases the soil acidity (*pH* from 5.7 to 7.3), iii) stimulates reduction
processes involving iron reducing bacteria indicated by the decrease of oxido reduction
potential *ORP* (84.6 to 9.2 mV) and the depletion of dissolved oxygen *DO* (1 mg/l to
less than 0.1 mg/l). Lime incorporation had also a significant impact on acidity

reduction (*pH* from 5.6 to 7.6). These results revealed that subsurface drainage implies different chemical processes as for the hematite soils of Tropical Savannah than the pyrite soils of mangroves and coastal flood plains. It alleviates soil acidity and will lead eventually to less ferrous iron intake by rice root, and thus improve the yields. Though lime achieves the same results, its difficult availability makes it a less competitive than drainage in tropical Savannah.

11. EVALUATION AND PERSPECTIVES

11.1. Scientific and engineering insights

11.1.1. Iron and clay stratification in some valley bottoms

Two statistical models – ANOVA and Welch T-test – were very helpful to analyse the properties of a unique soil, or to compare different soils in valley bottoms. ANOVA (Spiegel et al. 2001) was applied when the samples of response variables were confirmed as coming from normally distributed populations and when the variances of those samples were more or less equal. Alternately, when the variances were too different and the populations have similar distributions, but not necessarily normal, the non parametric Welch t-Test (Welch 1947) was applied. When significant differences between response variables' means were revealed by the preview tests, Fisher multi-test or Tukey pairwise comparisons were applied to find out which variables are equivalent and which are not. In this research project, the response variables have been as various as ferrous iron concentrations Fe^{2+}, clay percentage *%Clay*, organic matter content *%MO*, etc. The application of ANOVA, Welch t-Test, Fisher comparison, and Tukey pairwise comparisons to variable measured on random soils extracts taken from the field or from experimental units led to clear statements about the similarity of dissimilarities of the variable levels.

The study realized about clay and ferrous iron distributions in Tiefora, processed with ANOVA and Welch t-test and followed by multiple comparisons, led to important theoretical and practical results. It appeared that under some specific conditions – among which evaporation was mentioned – ferrous iron Fe^{2+} would precipitate essentially in the top soil – 30 cm top in the case of Tiefora – reaching values around 1000 mg/l or more. Similar distribution was found for clay in this valley bottom: in the top 100 cm, *%Clay* was found significantly higher – with a value of 30% – than underneath, thus making this zone a privileged area for intense chemical and microbiological activities. Such concentrations of clay and ferrous iron were not found in Moussodougou, where also iron toxicity is much more severe. Apparently, rice roots suffer less when ferrous iron is on top, above the majority of the rootlets, than when it is more uniformly distributed in the first 100 cm of top soil like in Moussodougou. These findings not only answer the *first research question* of this project, but also lead to interesting practical implications. To avoid excessive iron concentrations in the soil profile, maintaining permanent wet conditions seems a requirement. These wet conditions must be dynamic and not stagnant. Flow within the rootzone must be maintained for example through the use of subsurface drainage pipes (leaching). This can help in rootzone aeration but also prevent the excessive activity of iron and sulphate reducing bacteria. This finding, finally, stresses the necessity of modernizing traditional single-season irrigated rice valley bottoms into double or multi-season irrigation schemes, contributing the same way to farmers welfare and poverty alleviation.

11.1.2. How hematite soils differ from pyrite soil with mangroves

The ANOVA model and geochemical analysis, combined with microbiological interaction, led to the necessity of a paradigm shift about Tropical Savannah valley bottom soils in our fight against iron toxicity. These soils, despite of their appearance, are hematite Fe_2O_3 (iron III oxide) and not pyrite FeS_2 (iron disulphide). As shown in the survey study and the microplot experiment, when exposed to air – through subsurface drainage for example – Tropical Savannah valley bottom soils, at the opposite to pyrite, tend to become less acidic – from *pH* from 5.7 to 7.3 in one season in the case of Moussodougou. Here resides the need for a paradigm shift, with important consequences. In the field of microbiological research, it is noticeable that so far most studies (Traore et al. 1982, Ouattara and Jacq 1992, Dianou et al. 1998) in Tropical Savannah were concentrated on sulphate reducing bacteria (SRB), more present in pyrite soils than in hematite. Even though these bacteria are often associated, they are not the same. Reorienting the effort on understanding more the metabolism and the activity of iron reducing bacteria (IRB) associated with tropical savannah valley bottom hematite dominant soils may lead to the proposal of efficient microbiological solution to iron toxicity. In the field of subsurface drainage, while the increase of *pH* led to less acidic soils, this basification has a cost: ferrous iron concentration tends to increase, thus worsening iron toxicity threat. In short run, this is not a good sign. However, in medium and long run (associated for example with a fallow period), it will lead to alleviating iron toxicity since most of the metallic cations are less soluble, and thus tend to be less absorbed by rice roots, when the *pH* increases. Therefore, our results show that subsurface drainage can be a viable solution to the problem of iron toxicity. This was the answer to our *fifth research question*.

11.1.3. Permeability and clay analyses by regression

Depending on whether they are subject to flood or not, and the direction of the flooding, valley bottoms will have different permeability distribution that can be investigated through comparative non-linear regression analysis. In the case of Tiefora, the valley is flooded from downstream to upstream, especially during the rainy season. Alluvial fine particles are transported and will be deposit after bigger particle settled – i.e. upstream the valley bottom. Hence, at the very opposite of the particle deposition process happening in the reservoir for example, the permeability (or hydraulic conductivity) was found in Tiefora higher downstream than upstream. It increases for upstream with 0.10 ± 0.10 cm/h to more than 20.00 ± 10.00 cm/hr. Comparative non linear regression, applied to in situ underphreatic permeability (Lefranc test) measurements led, to a statistical confirmation of this interpretation. Hence, if any physical similar processes could be modelled by a particular type of regression equation, having the same number of coefficients statistically acceptable for all, a comparison of the coefficients would make it possible to conclude whether one process is slower, less intensive, lower etc. than another. In the case of Moussodougou, where the flood proceeds from the river banks and expands in the valley (spate irrigation), such a study was not performed. However, based on what was just said, it is reasonable to expect that the permeability will be higher near the river banks than further in the valley sides receiving the finest particles. Therefore, as engineering consequence, subsurface drainage pipes will be buried with a smaller spacing where the permeability is higher: downstream in Tiefora, and near the river in Moussodougou. Taking that into account may lead to important savings during a subsurface drainage project implementation in valley bottoms. These findings answered to the *third research question*.

Clay distribution through a valley toposequence may display some specific patterns calling for adapted drainage techniques. As shown in the case study of Tiefora, this pattern can be revealed by comparative non-linear regression. The shape of the plotted regression equation may provide precious information about how clay is spread over and underground and thus guide as to what drainage technique to apply. If a unique equation for the whole valley was found, a unique drainage technique can be applied, otherwise, the technique may differ from one location to another. In the case of Tiefora, from upstream to the middle section of the valley, a thick 2 m layer made of 20%-30% of clay persists on top, and mole drainage was proposed. In the downstream areas of the valley, clay proportion can rise to 40% but with more reduced thickness of 0.50 m and higher permeability, allowing the use of a more classical perforated pipe drainage technique (Hooghoudt method). These results answered the *second research question*.

11.1.4. Water use analysis: critical for irrigation improvement

ARIMA applied to a simple water level diver data records was found a very useful model to assess farmers' water use and propose better water management practices. Water use in most of the surface irrigation systems in Tropical Savannah is a cyclic time series process, with a non-seasonal – irrigation turn within the week – and seasonal components – water withdrawals adapted to rainy and dry seasons. Therefore Autoregressive Moving Average (ARIMA), which possesses the capacity to detect the trends in water use and to isolate the cycle span that ensures equity in the water distribution programme, can help tracking and evaluating irrigation practice month after month. This model is often used with financial and hydrological data. It was, during this project, successfully applied to the case of the 16 ha – irrigation scheme of Tiefora. Two main results were found: i) there is water waste, with a discharge of 200 m3/day permanently flowing when there is no irrigation because the gate is not closed; ii) the irrigation cycle that ensures equity in water distribution was not 5 days as planned by the farmers, but 20 days. The first finding called for a rehabilitation and/or modernisation of the sluices in the irrigation network, most of which were broken down or missing. The second finding called for the recognition of the 20 days as basis for equity in water distribution so as to defuse potential conflicts among satisfied and non-satisfied farmers. Answering to the *forth research question*, it also pointed to the need for better water management rules.

The capability of ARIMA applied to time series flow records at a main gate of a gravity system provides an important potential to follow up and improve water management in surface irrigation. In the case of Burkina Faso, more than 2000 small earthen dams exist and many are associated downstream with a surface irrigation system. However, the water use follow-up remains the great handicap. When water levels – which are afterwards converted into flows - are recorded, it is often done manually, by an operator. Hence, incomplete, irregular or erroneous data are frequently provided. When time series data exist in the national institutions, they are not processed for various reasons: the bad quality of the data, the lack of motivation of the agent, the lack of qualification, or the missing of processing tools. As a result, no water use or management is performed. This observation would be valid for most of Sub-Saharan African countries. There is a crucial lack of decisions based on a systematic data analysis. As important outcome, this project suggests the installation of automatic water level recording divers on a statistically and carefully selected set of surface irrigation schemes, because of their presumed waste of water. The data, extracted every three months for example, would be analysed and serve as basis to orient decisions and rehabilitation/modernisation operations. These activities, started within the framework of a 5-year project, must be handed over to a national institution responsible for water and agriculture.

11.1.3. Microplot experiments: efficient to track soil response

Microplot designed experiments combined with a processing through ANOVA General Linear Model (ANOVA GLM) help to isolate the key factors to alleviate or eradicate iron toxicity in valley bottom rice fields. Several factors such as microbial stems, type of soil, fertility, drainage conditions, soil acidity or rice cultivar with many others lurking factors influence simultaneously ferrous iron absorption by the rice roots. In a real rice growing field, lurking variable often acts differently upon the believed exact replicate farm plots, essentially because of the heterogeneity of the soil, but also due to irrigation and drainage operation variability from day to day. Exact replications are very difficult, if not unrealistic, in a real field. Several experiments led to contradictory results in such conditions. Nevertheless, replication is the master key to eliminate the distorting effect of lurking factors on the measured responses in the soil. The only way to obtain rigorous – though not perfect – replicates is through statistically well designed experiments. If a unique type of soil is introduced in a microplot and its exact copies in other identical microplots (its replicates), it is expected that any lurking factor will influence in the same way those replicates. Hence, if any little differential impact of this factor operates, averaging the responses of the replicates will produce small variances. Hence, a comparison between different factors' effect on the response can be performed through ANOVA GLM – which takes into account the variations around the means – and will yield statistically irreproachable results, including the potential interaction between the factors. One of the merits of this research project was to design and develop concrete microplots endowed with drainage and subsurface drainage capabilities. Research activities will be pursued with these microplots to yield, besides subsurface drainage, the fastest and most durable factors to alleviate rice iron toxicity.

11.2.   Socio economic impact of the research project

In order provide a feedback on the research results about iron toxicity, a research team went to the two Tropical Savannah irrigated rice valley bottoms of Tiefora (16 ha), and Moussodougou (36 ha) in the region of Comoé, located some 500 km away from the experiments site, in Burkina Faso. The main objective of this visit was to hold a meeting with each of the two farmers' organisations in order to inform them about what was found, propose solutions and have their thinking about the proposals.

11.2.1. Tiefora

In view of the critical importance of the historical records of water consumption on the irrigation valley bottom scheme, the team began to remind the farmers why the "E-water level" diver had been recording since 2013 the irrigation water consumption. So far, it is extremely rare - mainly because of the non-perception of the value of such monitoring – to get a time series of water consumption over several cropping seasons in surface irrigation systems. Yet such data provide powerful means of diagnosis and management, over time, of surface irrigation accused of being the main source of water losses in the production process. Concerns for diagnosing water consumption in Tiefora (16 ha), and eventually establish links with iron toxicity and low rice yield (less than 4 tons/ha), had led to the installation of this E-water level diver at the beginning of the primary canal. The team extracted the data recorded in the probe for the period of January 2013 to February to May 2014. Statistical analyses were performed on these data.

Afterwards, the team reported the surprising results obtained in rice yields of FKR19, an IN.ERA cultivar. This rice was grown on land taken from Tiefora and placed in 12 microplots (six drained, six non-drained) with 1 m² surface on each. The microplots were located at Kamboinsé, 15 km in the north of Ouagadougou. The experiments with these microplots yielded: 15 tons/ha during the first rainy season in 2011 and 16 tons/ha in the dry season of 2011/2012. We pointed out the strict application – in the process of production - standards IN.ERA fertilizer (split applications of 200 kg/ha of NPK 14-23-14 and 350 kg/ha of urea) in microplots – which are also rather confined environments in comparison to farmers' fields. High concentrations (950 mg/l) of Fe^{2+} ferrous iron were found in the 30 cm top soil in Tiefora during measurements in the field. Despite of finding similar values in our experiment in Kamboinsé, we could not observe any symptom of iron toxicity or any decrease in rice yield. However, field observations indicate the presence of precipitates of red iron (III) hydroxide in at least 90% of the plots, with traces of brown colour of iron on the leaves and grains of rice. These observations also show that if iron toxicity was present, the FKR 19 does not seem too suffer from it. Observations were also reported about aging nurseries, more than 30 days, regularly seen in farm plots in Tiefora, with major hindrance it poses to the tillering of rice. Furthermore, measurements on a network of nine piezometers installed in the valley show that the permeability is higher downstream (> 20 mm/h) than in the middle cross section (3.0 mm/h), which itself is more permeable than upstream (1.0 mm/h).

Thereafter, an open discussion took place between the team and the farmers, and some key recommendations were enacted to improve rice yield. (Figure 11.1-A). Farmers expressed their complete adherence to the observations brought to their attention. The research team then gave three recommendations to farmers. First, it was necessary to observe a strict compliance with the age of the nursery that is 21 days according the national agronomic research institute IN.ERA. Then, it was essential to apply the complex fertilizer NPK 14-23-14 as IN.ERA standards, but in combination with carefully making bunds around the plots leaving no room for water leakage into the surrounding farms. This could increase the confinement and the availability of fertilizer for rice. Finally, it would be critical to make gradual but steady incorporation of organic manure to reduce infiltration into areas of middle and downstream of the valley bottom.

Also in the permanent effort to support the farmers' organization in improving the management and production of rice in Tiefora, the research team provided three main documents to them (Figure 11.1-B). The first document is an aerial photo of the 500,000 m³ reservoir, including the village of Tiefora, the roads, the transport canal with its gallery forest and finally a GIS overlay of the detailed map of valley bottom irrigation scheme. This first map, explained team 2IE, is primarily used as a guide for quick orientation in group discussions with potential development partners. The second document was a map restoring the topography of the valley bottom as an overlay of the detailed farm plots map. It would be useful for the farmers, but especially for the technicians who wish to intervene in the valley for a few renovations/upgrades of equipments. Finally, the third level was the detailed plan, with an indication of the areas and parcel numbers. The last and third document was a map submitted with the current list of farmers in the valley bottom could serve at board meetings of their organisation and/or for the management of fees, fertilizers and pesticides distribution. The governing body farmers' organisation, on behalf of the group, expressed their gratitude for the provision of these maps and nominative lists (Figure 11.1-C).

11.2.2. Moussodougou

A few days after the visit in Tiefora, the research team travelled to Moussodougou, located some 75 km from Tiefora for a feedback meeting whose primary purpose was the ardent wish to report (Figure 11.2-A) to farmers the results of production experiments performed with their soil. These experiments were, as for Tiefora, also conducted in 12 microplots (6 drained, 6 non-drained) of 1 m² surface and 1 m³ of volume. In 2013, some 25 m³ of soil, whose locations were selected with the assistance of the farmers as highly contaminated ferrous iron soils were extracted with their collaboration from the 36 ha irrigated rice valley bottom of Moussodougou.. The rice yield on these remained horribly low (less than 1 tonne/ha) despite of an intensive incorporation of organic matter. A floating high concentration in ferric hydroxide was visible at the surface of the ponding water in farm plots. This soil was brought to the research experimental site of Kamboinsé near Ouagadougou, the Capital of Burkina Faso. On this site, tests were conducted during the 2014 season with the rice cultivar FKR19 without introduction of NPK fertilizer, using this organic soil of Moussodougou (organic material: 2 - 12% over the first 100 cm of soil). The results confirmed the statements of the farmers and the team reported that to them, with supporting photographs. Indeed, the team reported that the browning of rice - symptom generally accepted iron toxicity on ferrous iron intoxicated soils – appeared only one month after transplanting in 5 of 12 microplots, with scores of 1-3 (IRRI scale: surfaces of older leaves are reddish-brown, orange or yellow). Throughout the growth of rice, this iron toxicity had appeared in the 12 independent microplots, culminating at a score of 7 (most of the leaves are discoloured-brown, orange or yellow - or inactive). Except a drained microplot that made a rice paddy yield of 1.8 t / ha, the other eleven (drained or not) all received a zero or almost zero yield. The *pH* rose from 5.7 to 7.3 from beginning to end of the production process, within the top 50 cm of surface soil. Furthermore, a closer theoretical analysis of chemical equations and microbiological activity of SRB (sulphate reducing bacteria) and IRB (iron-reducing bacteria) that we conducted showed that iron toxicity is exacerbated in these hematite based soils by the organic matter inputs, incorporations often performed by the farmers of Moussodougou. The same theoretical analysis showed that the long dry season of more than six months during which almost nothing is grown in the valley, if it reduces the concentration of Fe^{2+} iron, also increases the acidity of the soil, making the rice more sensitive to iron toxicity for the next rainy season (Keïta et al. 2013b). Indeed ferrous iron is more soluble and therefore more easily absorbed by the rice roots in a highly acidic environment.

The report was followed by fruitful exchanges and supported advice to farmers. In fact farmers did confirm, among others, the worsening of iron toxicity when they incorporate organic manure to the soil. Nevertheless, given the importance of organic fertilizer for rice itself, the research team advised simply to moderate the use, without removing it completely. Conversely, based on the results of Tiefora, the team urged the use of NPK fertilizers in compliance with the standards of IN.ERA, and the careful erection of clayey bunds in order to confine the fertilizer within the farm plot and make it available for rice. The team also reported on its ongoing development of a detailed design, offering a project of sprinkler irrigation for Moussodougou in the dry season for vegetable crop growing. The project will use the shallow groundwater by making three tubewells, given that in this area flow rates as high as 20 m³/h could be expected. In the same line of action and as a continuation of our research activities on the site, some plots will be selected for subsurface drainage implementation to study their real impact on the alleviation of iron toxicity.

Finally, as for Tiefora, the research team also handed all three newly made maps (Figure 11.2-A) and the list of farmers in the organisation of Moussodougou. The documents were composed of the aerial photo map, the topographical map, and the farm plots system along with the list of farmers (see Appendix E).The latter included the ID numbers of the farm plots, the farm areas and indications of the presence of iron toxicity in the plot or not. It appeared again that 90% (i.e. some 32 ha) of the 36 ha were severely affected (much worse than Tiefora, with sometimes more than 3000 mg/l of Fe^{2+} in places) by iron toxicity. These documents, hitherto non-existent at Moussodougou, were greatly appreciated by the farmers (Figure 11.2- B, C). They promised to make excellent use for the management of the scheme and the fight against iron toxicity. They also greatly desired the continuation of the collaboration with our research team.

Figure 11.1: Restitution at Tiefora with the famers' association

(A) Plenary restitution with 25 farmers, Tiefora; (B) Detailed explanation of the new maps of the irrigation system with indication of iron intoxicated plots; (C) Photo with farmers' association, Tiefora

11.2.3. Social media

In order to contribute to knowledge improvement about investigating and developing land and water schemes in rural areas, the current research project has produced and uploaded on YouTube some 15 tutorial movies. The totality of these movies is available on YouTube. The area covered by these productions ranges from hydrometrics to microbiology, passing across geochemistry and small scale water saving irrigation equipment assembling. Several of these videos have been very appreciated by the public. For example, the video about "innovative irrigation systems in Sub-Saharan Africa (French)" has been viewed/downloaded some 500 times/month. Although reporting the results of a previous research-development project that involved the author, the broadcasting of this video was motivated by the current research dynamic to publish as much as possible results that the society may benefit from. The video "How to take a disturbed or undisturbed soil sample in a submerged soil at various depths (English)", though viewed/downloaded at a smaller rate of 45 times/month, was as well as the previous one classified by Google as "Creative Common". That meant that due to its potential interest for a wide public, a reuse of part or the complete video by third parties is authorised. Therefore, these videos and several others (Table 11.1) are expected to make a substantial contribution to land and water development sector in the coming years.

Figure 11.2: Restitution at Moussodougou with farmers' association

(A) Plenary restitution with 20 farmers, Moussodougou; (B) Detailed explanation of the maps with indication of iron inxoticated plots; (C) Photo with farmer's association, Moussodougou

Table 11.1: Some helpful videos created and uploaded to social media YouTube

Assembling manually innovative irrigation systems	French	14:55	https://www.youtu be.com/watch?v=d 9tklxPPzTw	*Creative Common*. 500 views/downloads per month
Extraction of soil samples	English	04:06	https://www.youtu be.com/watch?v=J YVGS9BLNTo	*Creative Common*. 45 views/downloads per month
How to use a colony counter	English	05:00	https://www.youtu be.com/watch?v=w UIZefiMXxI	130 views/downloads per month
Tutorial of using the RQFlex Plus 10 to measure ion concentrations in soils	French	02:22	https://www.youtu be.com/watch?v=x H8llSJ6CZU	50 views/downloads per month
Installation of a diver for time series water level recording in irrigation canal.	English	14:37	https://www.youtu be.com/watch?v=q 9Ds77Gn0qY	-

12. REFERENCES

(FAO) UNFaAO. 2013. Rice Market Monitor. Rome, Italy: Food and Agriculture Organization of the United Nations. 38 p.

Achim D, Thomas F. 2000. Rice: nutrient desorders and nutrient management. Potash & Phosphate Institute (PPI), Potash & Phosphate Institute of Canada (PPIC) and International Rice Research Institute (IRRI). 201 p.

Agung IGN. 2009. Time series data analysis using EViews. Singapore: John Wiley & Sons. 635 p.

Ahmad AR, Nye P. 1990. Coupled diffusion and oxidation of ferous iron in soils. I. Kinetics of oxygenation of ferrous iron in soil suspension. Journal of soil science. 41(395-409).

Allen RG, Pereira LS, Raes D, Smith M. 1998. Crop evapotranspiration - Guidelines for computing crop water requirements. Rome: FAO. 326 p.

Alvarez-Benedí J, Muñoz-Carpena R (eds.). 2005. Soil-water-solute process characterization. An integrated approach. USA: CRC Press. 816 p.

Anderson LR. 1941. Distribution of the serial correlation coefficients. Annals of Math. Statistics. 8(1):1-13.

Andreisse W, Fresco LO. 1991. A charcterisation of rice growing environments in West Africa. Elsevier Science Publishers B.V. 33:377-395.

Antony J. 2003. Design of experiments for engineers and scientists. Burlington. UK: Butterworth Heinemann/Elsevier. 165 p.

Arth I, Frenzel P, Conrad R. 1998. Denitrification coupled to nitrification in the rhizosphere of rice Soil biology. Biochemistry. 30:509-515.

Attanandana T, Vacharotayan S. 1986. Acid Sufate Soils: Their Characteristics, Genesis, Amelioration and Utilization. Southeast Asian Studies. 24(2):154-180.

Audebert A, Sahrawat KL. 2000. Mechanisms for iron toxicity tolerancein lowland rice. Journal of plant nutrition. 23(11 & 12):1877-1885.

Bajwa MS, Josan AS, Hira GS, Singh NT. 1986. Effect of sustained saline irrigation on soil salinity and crop yields. Irrigation Science. 7(1):27-35.

Balley RA. 2008. Design of comparative experiments. Cambridge University Press. 346 p.

Barron V, Torrent J. 1986. Use of the Kubelka-Munk theory to study the influence of iron oxides on soil colour. Journal of Soil Science. 37:499-510.

Becker M, Asch F. 2005. Iron toxicity in rice - conditions and management concepts. J. Plant Nurti. Soil Sci. 168:558-573.

Becker M, Johnson DE. 1999. Rice yield and productivity gaps in irrigated systems of the forest zone of Côte d'Ivoire. Field Crops Research. 60(1999):201-208.

Becker M, Johnson DE. 2001. Improved water control and crop management effects on lowland rice productivity in West Africa. Nutrient cycling in agroecosystems. 59:119-127.

Begg CBM, Kirk GJD, Mackenzie AF, Neuc II-U. 1994. Root-induced iron oxidation and pH changes in the lowland rice rhizosphere. New phytology. 128:469-477.

Bel Ombre Development Company (BODCOLTD). 2011. Sikalite [Online]. Available: https://www.youtube.com/watch?v=GB2xsJ1-A1c 18/08/2014.

Bender M, Simonovic S. 1994. Time-Series Modeling for Long-Range Stream-Flow Forecasting. Journal of Water Resources Planning and Management. 120(6):857-870.

Berthelin J, Boymond D. 1978. Some aspects of the role of heterotrophic microorganisms in the degradation of minerals in waterlogged acid soils. Environmental biochemistry and geomicrobiology. 2:659-673.

Berthelin J, Kogblevi A. 1974. Inlfuence de l'engorgement sur l'altération microbienne des minéraux dans les sols. Revue écologie, biologie et sols. 11:499-509.

Bier AW, Lange H. 2009. Introduction to Oxidation Potential Measurement. In: Hach Company (ed.). USA: Hach Company, .

Birmani SS, Mao CX, Hardy B (eds.). 2003. Hybrid Rice for Food Security, Poverty Alleviation, and Environmental Protection. Manila, Philippines: International Rice Research Institute. 404 p.

Boslaugh S, Watters PA. 2008. Statistics in a Nutshell. USA: OReilly. 478 p.

Boutilier RG, Shelton G. 1980. The Statistical Treatment of Hydrogen Ion Concentration And pH. Journal of Experimental Biology. 84:335-339.

Bouwer H, Jackson RD. 1974. Determining Soil Properties. In: Drainage for agriculture. Shilfgarde van J. (ed.). p. 611-672.

Box GEP, Jenkins GM, Reinsel GC. 1994. Time series analysis forecasting and control. In: Riker E (ed.). New Jersey, USA: Prentice Hall International.

Brase CH, Brase CP. 2007. Understanding Basic Statistics. USA: Houghton Mifflin Company. 542 p.

Breemen NV. 1992. Environmental Aspects of Acid Sulphate Soils. In: Dent DL & Mensvoort MEFv, eds.International Symposium on Acid Sulphate Soils; 1992. Ho Chi Minh City, Vietnam. ILRI.

Bronkhorst R. 2006. Effets de l'aide alimentaire structurelle dans la forme d'achat local et de vente de riz sur le développement rural. Etude de cas sur le Burkina Faso [Effects of food structural assistance in term of buying and selling local rice on rural development. Case study of Burkina Faso]. Ouagadougou (Burkina Faso): Rapport au Ministère des Affaires Etrangères des Pays-Bas (DGIS/DDE/IM). 46 p.French.

Butlin KR, Adams ME, Thomas M. 1949. The Isolation and Cultivation of Sulphate-Reducing Bacteria. Microbiology. 3(1):46-59.

Caires EF, Garbuio FJ, Churka S, Barth G, Corrêa JCL. 2007. Effects of soil acidity amelioration by surface liming on no-till corn, soybean, and wheat root growth and yield. European Journal of Agronomy. 28:57-64.

Cam C, Froger D, Moulin J, Rassineux J, Servant J. 1996. Representation cartographique de la sensibilité des sols à l'infiltration hydrique verticale. (French) [cartographic representation of the soils ability to vertical infiltration - thematic map of vertical infiltration]. Etudes et Gestion des Sols 3:97-112.

Cannell RQ, Goss MJ, Harris GL, Jarvis MG, Douglas JT, Howse KR, Grice SL. 1984. A study of mole drainage with simplified cultivation for autumn-sown crops on a clay soil. The Journal of Agricultural Science. 102(3):539-559.

Carroll RJ, Schneider H. 1985. A note on levene's tests for equality of variances. Statistics & Probability Letters. 3(4): 191–194.

Cassan M. 2005. Les essais de perméabilité sur site dans la reconnaissance des sols [Permeability in-situ tests for the indentification of soils]. Paris. 561 p.French.

Charles H, Godfray J, Beddington JR, Crute IR, Haddad L, David Lawrence, Muir JF, Pretty J, Robinson S, Thomas SM, Toulmin C. 2010. Food Security: The Challenge of Feeding 9 Billion People. Science. 327:812-818.

Chaudhry MH. 2008. Open-Channel Flow. 2nd ed. New York, USA: Springer.

Chérif M, Audebert A, Fofana M, Zouzou M. 2009. Evaluation of Iron Toxicity on Lowland Irrigated Rice in West Africa TROPICULTURA. 27(2):88-92.

Chesworth W (ed.) 2008. Encyclopedia of soil science. Dordrecht, The Netherlands: Springer. 860 p.

Christen EW, Ayars JE. 2001. Subsurface drainage system design and management in irrigated agriculture: best management practices for reducing drainage volume and salt load. Irrigation Science. 21:35-43.

Clarke FW. 1924. Data of Geochemistry Washington (D.C.): U.S. Geological Survey Bull. 770 p.

Coles ED. 1968. Some Notes on Drainage Design Procedure. Proceedings of The South African Sugar Technologists' Association. Manzini, Swaziland. 189-199.

Comité Inter-Africain d'Etudes Hydrauliques (CIEH). 1987. Recherche sur la fracturation profonde en zone de socle cristallin à partir de forages à gros débit et de linéaments Landsat à l'aide de méthodes géophysiques avancées. Ouagadougou [Research of deep fractures in the cristalline basement area using boreholes logs with important discharge and the lineation of Lansat with advanced geophysical methods]. Ouagadougou (Burkina Faso). French.

Conklin Jr. AR. 2005. Introduction to Soil Chemistry Analysis and Instrumentation. Vol.167. USA: Wiley and Sons Inc. 241 p.

Crowe PR. 1954. The effectiveness of precipitation. A graphical analysis of Thornthwaite's climatic classification. Geographical Studies. 1(1):44-62.

Dalton TJ, Guei RG. 2003. Productivity grains from rice genetic enhancements in West Africa : countries and ecologies. Elsevier Science Direct. 31(2):359-374.

Dave G. 1985. The Influence of pH on the Toxicity of Aluminum, Cadmium, and Iron to Eggs and Larvae of the Zebrafish, *Brachydanio rerio* Ecotoxicology and Environmental Safety. 10:253-267.

de Oliveira EL, Pavan MA. 1996. Control of soil acidity in no-tillage system for soybean production. Soil & Tillage Research. 38:37-47.

de Veer M, Wormgoor JA, Rizq RG, Wolters W. 1993. Water management in tertiary units in the Fayoum, Egypt. Irrigation and drainage systems. 7(1):69-82.

DeBont JAM, Lee KK, Bouldin DF. 1978. Bacterial oxidation of methane in a rice paddy. Ecology Bulletin. 26:91-96.

Dembele Y, Yacouba H, Keïta A, Sally H. 2012. Assessment of irrigation system performance in south-western Burkina Faso. Irrigation and Drainage. 61(3):306-315.

Dent D. 1986. Acid sulphate soils: a baseline for research and development. Vol.39. Wageningen, The Netherlands: International Institute for Land Reclamation and Improvement/ILRI. 24 p.

Dianou D. 1993. Etude de l'influence des interactions entre les bactéries réductrices du cycle du soufre et les bactéries méthanigènes sur la production du riz des bas-fonds [Study of the influence of the interactions between sulfate reducing bacteria and methogenic bacteria on valley bottom rice production]. PhD. Ouagadougou: Univesité de Ouagadougou. French.

Dianou D. 2005. Etude des bactéries sulfato-réductrices, méthanotrophes et des archaeobacteries méthanogènes dans les sols de rizière : influences des interactions sur la production du riz et sur l'émission de méthane dans l'atmosphère [Study of the Sulfate Reducing Bacteria, the Methanotroph and the Methanogenic Archaeobacteria in Rice Growing Soils: Influence of the

Interactions on Rice Production and the Methane Emission in the Atmosphere]
PhD. Ouagadougou: University of Ouagadougou. French.

Dianou D, Lopes J, Traoré AS, Lino A-R, Moura I, Moura JG. 1998. Characterization of Desulfovibrio sp. isolated from some lowland paddy field soils of Burkina Faso. Soil Science and Plant Nutrition. 44(3):459-465.

Dorosh PA. 2001. Trade Liberalization and National Food Security: Rice Trade between Bangladesh and India. World Development. 29(4):673-689.

Easterly W. 2009. How the Millennium Developement Goals are unfair to Africa. Elsevier. 37(1):26-35.

El-Swaify SA, Emerson WW. 1975. Changes in the physical properties of soil clays due to precipitated aluminum and iron hydroxides. I. Swelling and aggregate stability after drying. Soil Science Society of America. 39:1056-1063.

Emerson D, Weiss JV, Megonigal JP. 1999. Iron-Oxidizing Bacteria Are Associated with Ferric Hydroxide Precipitates (Fe-Plaque) on the Roots of Wetland Plants. Applied and Environmental Microbiology. 65(6):2758-2761.

Fagbamia A, Ajayia FO. 1990. Valley Bottom Soils of The Sub-Humid Tropical Southwestern Nigeria on Basement Complex: Characteristics and Classification. Soil Science and Plant Nutrition. 36(2):179-194.

Fairhurst TH, Dobermann A. 2002. Rice in the Global Food Supply. Better Crops International. 16, Special Supplement:1-4.

Feddema J. 2005. A revised thornthwaite type global climate classification. Physical Geography. 26(6):442 - 466.

Fisher WR, Schwertmann U. 1974. The Formation of Hematite from Amourphous Iron(III) Hydroxide. Clays And Clay Minerals. 23:33-37.

Foth HD. 1990. Fondamentals of soil science, 8th ed. USA: John Wiley & Sons. 382 p.

Fredrickson JK, Gorby YA. 1996. Environmental Processes Mediated by Iron-reducing Bacteria. Current Opinion in Biotechnology. 7:287-294.

Freney JR, Jacq VA, Baldensperger JF. 1982. The significance of the bilogical sulfur cycle in rice production. Microbilogy of tropical soils and plant productivity.271-317.

Frenken K. 2005. Irrigation in Africa in Figures: Aquastat survey 2005. Food and Agriculture Organization (FAO). 74 p.

Gardner DG, Gardner JC, Laush G, Meinke WW. 1959. Method for the Analysis of Multicomponent Exponential Decay Curves. The Journal of Chemical Physics. 31(4):978-986.

Gastwirth JL, Gel YR, Miao W. 2009. The impact of Levene's Test of Equality of Variances on statistical theory and practice. Statistical Science. 24(3):343-360.

Gerard G, Chanton J. 1993. Quantification of methane oxidation in the rhizosphere of emergent aquatic marocphytes - defining upper limits. Biogeochemistry. 23:79-97.

Gilbert B, Frenzel P. 1995. Methanotrophic bacteria in the rhizosphere of rice microcosms and their effec on pore water methane concenctration and methane emission. Biology. Fertilisation. Soils. 20:93-100.

Griffiths JF. 1983. Climatic classification. International journal of environmental studies. 20(2):115-125.

Hamilton JD. 1994. Time series analysis. New Jersey, USA: Princeton University Press.

Härdle W, Mammen E. 1993. Comparing nonparametric versus parametric regression fits. The Annals of Statistics. 21(4):1926-1947.

Haynes RJ, Naidu R. 1998. Influence of lime, fertilizer and manure applications on soil organic matter content and soil physical conditions: a review. Nutrient Cycling in Agroecosystems. 51:123-137.

Henderson FM. 1966. Open channel flow. London, UK: MacMillan Publishing Co., Inc.

Herzsprung P, Friese K, Packroff G, Schimmele M, Wendt-Potthoff K, Winkler M. 1998. Vertical and annual distribution of ferric and ferrous iron in acidic mining lakes. Acta hydrochimica et hydrobiologica. 26(5):253-262.

Hillel D. 2004. Introduction to Environmental Soil Physics. California (USA). 511 p.

Hillel D, Rosenzweig C, Powlson D, Scow K, Singer M, Sparks D (eds.). 2004. Encyclopedia of soils in the envrironment. Vol.3. New York (USA): Acdemic Press. 613 p.

Ho DD, Neumann AU, Perelson AS, Chen W, Leonard J, Markowitz M. 1995. Rapid Turnover of Plasma Virions and CD4 lymphocites in HIVE-1 Infection. Nature. 373(12):123-126.

Hoaglin DC, Mosteller F, Tukey JW (eds.). 1991. Fundamentals of exploratory analysis of variance. USA: John Wiley & Sons. 456 p.

Holzapfel-Pschorn A, Conrad R, Seiler W. 1985. Production, oxidation and emission of methane in rice paddies. FEMS Microbiology. Ecology. 31:343-351.

Horton RE. 1941. An approach towards physical interpretation of infiltration capacity, soil science society of american. Soil Science Society of America Journal. 5(C):399-417.

Housecroft CE, Constable EC. 2006. Chemistry. Harlow, England: Pearson Education Limited. 1317 p.

Hussain I, Hanjra MA. 2003. Does irrigation water matter for rural poverty alleviation ? Evidence from South and South-East Asia. Water Policy. 5(2003):429-442.

Institut de l'Environnement et de Recherches Agricoles (IN.ERA). 2000. Fiches descriptives des variétés de riz. (French) [Descriptive folders of rice cultivars]. Ouagadougou (Burkna Faso): Institute of Environment and Agricultural Reasearch. 56 p.French.

Institut National de la Statistique et de la Démographie (INSD-BF). 1985. Recensement général de la population. Structure par âge et sexe des villages du Burkina Faso. [General Sensus of The Population. Structure per Age and Sex of the Villages of Burkina Faso]. Ouagadougou (Burkina Faso). French.

Institut National de la Statistique et de la Démographie (INSD-BF). 2011. Indices du commerce extérieur du Burkina Faso [Indices of External Trade of Burkina Faso]. Ouagadougou (Burkina Faso). 30 p.French.

International Rice Research Institute (IRRI). 1985. Soil physics and rice. Los Baños Laguna, Philippine: International Rice Research Institute (IRRI). 436 p.

International Rice Research Institute (IRRI). 2002. Standard Evaluation Systems for Rice. Philippines: International Rice Research Institute. 56 p.

Inubishi K. H. 1984. Easily decomposable organic matter in paddy soils. IV. Relationship between reduction process and organic matter decoposition. Soil science. Plant nutirion. 30:189-198.

Jackson ML, Sherman GD. 1953. Chemical Weathering of Minerals in Soils. Advances in agronomy. 5:219-318.

Jacq VA. 1989. Participation des bactéries sulfo-réductrices aux processus microbiens de certaines maladies physiologiques du riz inondé (exemple du Sénégal). PhD thesis. [Participation of sulfate reducing bacteria in microbial processes of some physiological disorders of flooded rice (example of Senegal)]. Marseille (France): Université d'Aix-Marseille I. France. French.

Jacq VA, Fortuner R. 1979. Biological control of rice nematods using sulphate reducing bacteria. Revue nématologique. 2(1):41-51.

Jacq VA, Ottow JCG. 1991. Iron Sulphide Accumulation in the Rhizosphere of Wetland Rice (*Oryza sativa L.*) as Result of Microbial Actitivies. In: Diversity of Environmental Biogeochemistry. Berthelin J (ed.). La Haye: Elsevier. p. 453-468.

Jacq VA, Prade K, Ottow JCG. 1988. Iron sulphide accumulation in the rhizosphere of wetland rice (oryza sativa L.) as the result of microbial activities. *Horizon documentation IRD.*

Jones MP, Dingkuhn M, Aluko GK, Mandé S. 1997. Interspecific Oriza Sativa L. X O. Glaberrima Steud. progenies in upland rice improvement. Euphytica - Kluver Academic Publishers. 92:237-246.

Kanté I. 2011. Caractérisation topographique, hydrologique et pédologique de l'aménagement irrigué de bas-fond de Tiéfora en aval de barrage dans la province de la Comoé [Caracterisation of the topography, hydrology and pedology of the valley bottom irrigation field of Tiéfora in the Comoe province]. MSc. Ouagadougou (Burkina Faso): International Institute for Water and Environmental Engineering (2iE). French.

Kato Y, Suzuki K, Nakamura K, Hickman AH, Nedachi M, Kusakabe M, Bevacqua DC, Ohmoto H. 2008. Hematite Formation by Oxigenated Groundwater More than 2.76 Billion Years Ago. Earth and Planetary Science Letters. 278(1-2):40-49.

Keïta A. 2013a. Ewater level installation Tiefora [Online]. Youtube. Available: https://www.youtube.com/watch?v=q9Ds77Gn0qY.

Keïta A. 2013b. Guide d'utilisation du réflectomètre RQFlex Plus 10 [The reflectometer user guide] [Online]. YouTube. Available: http://www.youtube.com/watch?v=xH8lISJ6CZU [Accessed 18/08/2013 2013].

Keïta A. 2014. How to take a disturbed or undisturbed soil sample in a submerged soil at various depths [Online]. Youtube. Available: https://www.youtube.com/watch?v=JYVGS9BLNTo.

Keïta A, Hayde LG, Yacouba H, Schultz B. 2014a. Valley bottom clay distribution and adapted drainage techniques. Journal of Lowland Technology International. 16(2):135-142.

Keïta A, Schultz B, Yacouba H, Hayde LG. 2013a. Clay and ferrous iron stratifications in a Tropical Savannah valley bottom soil under irrigated rice. Academia Journal of Agricultural Research. 1(11):204-210.

Keïta A, Yacouba H, Hayde LG, Schultz B. 2013b. A single-season irrigated rice soil presents higher iron toxicity risk in Tropical Savannah valley bottoms. Open Journal of Soil Science. 3:314-322.

Keïta A, Yacouba H, Hayde LG, Schultz B. 2014b. Assessing irrigation water management using trend analysis and autocorrelation International Agricultural Engineering Journal. In Press.

Keïta A, Yacouba H, Hayde LG, Schultz B. 2014c. Comparative non-linear regression - a case of infiltration rate increase from upstream in valley. International Agrophysics. 28:303-310.

Kessler J, Oosterbaan RJ. 1974. Determining hydraulic conductivity of soils. In: Public. I, ed.Drainage principles and application; 1974. Wageninger, the Netherlands. ILRI. 255-295.

King GM. 1994. Association of methanotrophs with the roots and rhizomesof aquatic vegetation. Applied environmental microbiology. 60:3220-3227.

Kirk GJD. 1995. Root-induced iron oxidation, pH changes and zinc solubilization in the rhizosphere of lowland rice. New phytology. 131:129-137.

Koppen W. 1936. Das geographische System der Klimate (German) [The geographical classification of climate]. Vol.I, Part C. Berlin (Germany): Gebruider Borntraeger. 44 p.German.

Kusky T. 2005. Encyclopedia of earth science. New York, USA: Facts of File. 529 p.

Lançon F, Benz HD. 2007. Rice Imports in West Africa: Trade Regimes and Food Policy Formulation. Pro-poor Development in Low Income Countries; 2007. France. EAAE.

Levene H. 1960. Robust tests for equality of variances. USA: Stangord University Press. 278-292 p.

Liesack W, Schnell S, Revsbech NP. 2000. Microbilogy of flooded rice paddies. FEMS Microbiology Reviews. 24:625-645.

Lotze-Campen H, Müller C, Bondeau A, Rost S, Popp A, Lucht W. 2008. Global food demand, productivity growth, and the scarcity of land and water resources: a spatially explicit mathematical programming approach. Agricultural Economics. 39:325-338.

Lovly DR. 1987. Organic matter mineralization with the reduction of ferric iron : a review. Geomicrobiology Journal. 5:375-399.

Majerus V, Bertin P, Lutts S. 2007. Effects of iron toxicity on osmotic potental, osmolytes and polyamines concentrations in the African rice (*Oriza glaberrima* Steud.). Elsevier Science Direct. 173:96-105.

Masiyandima MC, van de Giesen N, Diatta S, Windmeijer PN, Steenhuis TS. 2003. The hydrology of inland valleys in the Sub-Humid Zone of West Africa: rainfall-runoff processes in the M'be experimental watershed. Hydrological Processes. 17(6):1213-1225.

Mason RL, Gunst RF, Hess JL. 2003. Statistical design and analysis of experiments with applications to engineering and science. 2nd.USA: Willey-Interscience. 752 p.

Materechera SA, Mkhabela T. 2002. The effectiveness of lime, chicken manure and leaf litter ash in ameliorating acidity in a soil previously under black wattle (Acacia mearnsii) plantation. Bioresource Technology. 85:9-16.

Materechera SA, Mkhabela TS, Antony J. 2002. The effectiveness of lime, chicken manure and leaf litter ash in ameliorating acidity in a soil previously under black wattle (Acacia mearnsii) plantation. Bioresource Technology. 85:9-16.

Mathew EK, Panda R, Nair M. 2001. Infuence of subsurface drainage on crop production and soil quality in a low-lying acid sulphate soil. Agricultural Water Management. 47:191-209.

Mathews PG. 2005. Design of experiments with Minitab. Wisconsin. USA: ASQ Quality Press. 521 p.

McWilliams TP. 1990. A distribution-free test for symmetry based on a runs statistic. Journal of the American Statistical Association. 85(412):1130-1133.

Miller W. 2013. Statistics and measurements concepts with OpenStat. New-York, USA: Springer.

Ministère de l'Administration Territorial et de la Décentralisation (MATD-BF). 2005. Répertoire des villages administratifs du Burkina Faso [Directory of villages of Burkina Faso]. Ouagadougou (Burkina Faso). French.

Ministère de l'Agriculture de l'Hydraulique et des Ressources Halieutiques (MAHRH-
    BF). 1999. Inventaire des bas-fonds de l'Ouest et du Sud-Ouest. Vol 1 :
    Rapport principal [Inventory of the West and South-Western Burkina Faso.
    Vol 1: Main report]. Ouagadougou (Burkina Faso): MAHRH. 126 p.French.

Montgomery DC. 2001. Design and analysis of experiments. 5th.New York, USA: John
    Wiley & Sons. 699 p.

Montgomery DC, Jennings CL, Kulahci M. 2008. Introduction to time series analysis
    and forecasting. In: Balding DJ, Cressie NAC, Fitzmaurice GM, Johnstone IM,
    Molengerghs G, Scott DW, Smith AFM, Tsay RS & Weisberg S (eds.) *Wiley
    Series in Probability and Statistics.* Hoboken, New Jersey: John Wiley & Sons
    Inc.

Montgomery DC, Runger GC. 2011. Applied Statistics and Probability for Engineers.
    Wiley and Sons, Inc. 792 p.

Moormann FR, Breemen NV. 1978. Rice: Soil, Water, Land. Los Banos Laguna
    (Philippines): International Rice Research Institute (IRRI).

Moreau DH, Pyatt EE. 1970. Weekly and monthly flows in synthetic hydrology. Water
    Resources Research. 6(1):53-61.

Murray MB. 1994. Environmental Chemistry of Soils. USA: Oxford University Press.
    411 p.

Nagano T, Nakashima S, Nakayama S, Osada K, Senoo M. 1992. Color Variations
    Associated With Rapid Formation Of Goethite From Proto-Ferrihydrite At Ph
    13 And 40°C. Clays and Clay Minerals. 40(5):600-607.

Namara RE, Hanjra MA, Castillo GE, Ravnborg HM, Smith L, Koppen BV. 2009.
    Agricultural water management and poverty linkages. Agricultural Water
    Management. 97(2010):520-527.

Neumeyer N, Dette H. 2003. Nonparametric comparison of regression curves: an
    empirical process approach. The Annals of Statistics. 31(3):880-920.

Nguu NV, Gibbons JW, Dobson RL. 1988. Performance of rice (Oryza sativa) on slopes
    of inland valleys in West Africa. Elsevier Science Publishers B.V. 18:113-125.

Nguyen NV, Ferrero A. 2006. Meeting the Challenges of Global Rice Production.
    Spinger-Verlag. 4:1-9.

Office National des Barrages et des Aménagements Hydroagricoles (ONBAH). 1987.
    Study of small dams and irrigation schemes in the South-West Feasibility of
    the site of Tiéfora. Ouagadougou (Burkina Faso). 21 p.French.

Ogban P, Babalola O. 2009. Characteristics, classification and management of inland
    valley bottom soils for crop production in sub-humid southwestern Nigeria.
    Journal of tropical agriculture, food, environment and extension. 8(1):1-13.

Oosterbaan RJ, Gunneweg HA, Huizing A. 1986. Water control for rice cultivation in
    small valleys of West Africa. Wageningen, The Netherlands: International
    Institute for Land Reclamation and Improvement (ILRI). 15 p.

Oster JD. 1994. Irrigation with poor quality water. Agricultural Water Management.
    25(1994):271-297.

Otoidobiga H. 2012. Sudy of the impact of sulphate and iron reducing bacteria on the
    vegetative development and the yields of flooded rice under subsurface
    drainage (French). MSc. Burkina Faso: University of Ouagadougou.

Ottow JCG, Benckiser G, Watanabe I. 1982. Toxicity of rice as a multiple nutritional
    soil stress. Tropical agriculture research. 15:167-179.

Ottow JCG, Fabig W. 1985. Influence of oxygen aeration on denitrification and redox
    level in different bacterial batch cultures. Planetary ecology.427-440.

Ottow JCG, Glathe H. 1971. Isolation And Identification Of Iron-Reducing Bacteria From Gley Soils. Soil Biology and Biochemistry. 3:43-55.

Ouattara AS, Jacq VA. 1992. Characterization of sulfate-reducing bacteria isolated from Senegal rice fields. ScienceDirect. 101(3):217-228.

Ouattara BS. 1992. Contribution to the study of iron and sulfate reducing bacteria in the Kou valley bottom rice field. PhD Microbiologie. Aix-Marseille: Université de Provence Aix-Marseille I. French.

Ouédraogo M, Dembélé Y, Dakouo D. 2005. The problem of commecialisation of paddy rice and adaptation strategies of farmers in large scale irrigation systems in the Western Burkina Faso (French). Atelier régional sur les politiques rizicoles et sécurité alimentaire en Afrique Sub-saharienne; 2005. Centre du riz pour l'Afrique (ADRAO), du 7 au 9 novembre 2005, Cotonou (Bénin). Ouagadougou, Burkina Faso. MESSRS, Burkina Faso. 17 p.

Patrick JWH. 1960. Nitrate reduction rates in a submerged soil as affected by redox potential. 7th International Congress. Soil Science. Madison (USA).

Peel MC, Finlayson BL, McMahon TA. 2007. Updated world map of the Koppen-Geiger climate classification. Hydrology and Earth System Sciences. 11:1633–1644.

Persson KB. 1997. Soil phosphate analysis: a new technique for measurement in the field using a test strip. Archaeometry. 39(2):441-443.

Peter DJ. 1990. Methods for measuring the saturated hydraulic conductivity of tills. Nordic Hydrology. 21:95-106.

Ponnamperuma FN. 1972. The chemistry of submerged soils. Advances in Agronomy. 24:29-96.

Pons LJ, Breemen NV, Driessen PM. 1982. Pysiography of coastal sediments and development of potential soil acidity. In: Kittrick JA, Fanning DS & Hossner LR (eds.) Acid Sulfate Weathering. USA: Soil Science Society of America.

Prade K, Ottow JCG, Jacq VA, Malouf G, Loyer J-Y. 1990. Relationship between soil properties of rice flooded soils and iron toxicity in Low Casamance (Senegal). Litterature review of previous studies (French). Cahiers ORSTOM. XXV(4):453-474. French.

Pretty JN, Morison JIL, Hine RE. 2003. Reducing food poverty by increasing agricultural sustainability in developing countries. Elsevier 95:217-234.

Raunet M. 1985. Bas-fonds et riziculture en Afrique. Approche structurale comparative. L'Agronomie Tropicale. 40(3):181-200.

Reddy KR, Patrick Jr WH, Lindau CW. 1989. Nitrification-denitrification at the plant root-sediment interface in wetlands. Limnology. Oceanography. 34:1004-1013.

Rickard D, Luther GW. 2007. Chemistry of Iron Sulfides. Chemistry Revisions. 107:514-562.

Ritzema HP, Satyanarayana TV, Raman, Boonstra J. 2008. Subsurface Drainage to Combat Waterlogging and Salinity in Irrigated Lands in India: Lessons Learned in Farmers' Fields. Agricultural Water Management. 95:179-189.

Robinson M, Mulqueen J, Burke W. 1987. On flows from a clay soil - Seasonal changes and the effect of mole drainage. Journal of Hydrology. 91(3-4):339-350.

Roger P. 1991. Les biofertilisants fixateurs d'azote en riziculture : potentialités et facteurs limitants (French). Biofertilizer nitogen fixers in rice growing : potentialities and limiting factors. In: Maunet R, ed.Séminaire international bas-fonds et riziculture, 9-14 décembre 1991; 1991. Madagascar. Montpellier Cedex, France. CIRAD CA. 3327-348.

Royal Eijkelkamp. 2009. Sample Ring Kits - Operating Instructions. ZG Giesbeek, The Netherlands: Royal Eijkelkamp.

Rumsey D. 2009. Statistics II for Dummies. USA: Wiley Pusblishing, Inc. 178 p.

Sahravat KL. 2004. Managing iron toxicity in lowland rice: the role of tolerant genotypes and plant nutrients. In: Rice is life: scientific perspectives for the 21st century. Toriyama K, Heong KL & Hardy B (eds.). Tsukuba, Japan. p. 452–454.

Sahravat KL, Mulbah CK, Diatta S, DeLaune RD, Patrick WH Jr, Singh BN, Hones MP. 1996. The role of tolerant genotypes and plant nutrients in the management of iron toxicity in lowland rice. J. Agric. Schi. Cambridge. 126:143-149.

Sahrawat K. 2004. Managing iron toxicity in lowland rice: the role of tolerant genotypes and plant nutrients. In: Toriyama K, Heong KL & Hardy B, eds.Rice is life: scientific perspectives for the 21st century. Tsukuba, Japan. 452–454.

Salas JD, Delleur JW, Yevjevich VM, Lane WL. 1980. Applied modeling of hydrologic time series. Littleton, Colorado, Water Resources Publications.

Scala DJ, Hacherl EL, Cowan R, Young LY, Kosson DS. 2006. Characterization of Fe(III)-reducing enrichment cultures and isolation of Fe(III)-reducing bacteria from the Savannah River site, South Carolina. Research in Microbiology. 157(8):772-783.

Schaetzl RJ, Sharon A. 2005. Soils genesis and geomorphology. New York (USA): Cambrige University Press. 833 p.

Schlichting E. 1973. Pseudogley and gley: genesis and use of hydromorphic soils. In: Schlichting E & Schwertmann U, eds. Weinheim, Germany, Chemie Verlag. Tansaction Commission V and VI of ISSS. 71-80.

Schlumberger. 2006. Product manual of Mini Diver, Macro Diver, Cera Diver and Baro Diver. Deflt, The Netherlands: Sclumberger water services. 27 p.

Schultz B. 1988. Drainage measures and soil ripening during the reclamation of the former sea bed in the IJsselmeerpolders. Proceedings 15th European regional conference of ICID on Agricultural Water Management; 1988. Dubrovnik (Yugoslavia). International Commission on Irrigation and Drainage (ICID). 197-214.

Schultz B, Tardieu H, Vidal A. 2009. Role of water management for global food production and poverty alleviation. Irrigation and Drainage. 58(S1):S3-S21.

Schwertmann U., Murad E. 1983. Effects of pH on the formation of goethite and hematite from ferryhidrite. Clays And Clay Minerals. 31(4):277-284.

Seckler D. 1996. The new era of water resources management. Colombo, Sri Lanka: International Irrigation Management Institute (IIMI). 12 p.

Selley RC. 2000. Applied Sedimentology 2nd Edition. California, USA: Academic Press. 543 p.

Shamway RH, Stoffer DS. 2011. Time series analysis and its applications, with R examples. 3rd.New York, USA: Springer. 609 p.

Shapiro S, Wilk MB. 1965. An analysis of variance test for normality: complete samples. Biometrika 52(3-4):591-611.

Sharma ML, Barron RJW, Fernie MS. 1987. Areal distribution of infiltration parameters and some soil physical properties in lateritic catchments. Journal of Hydrology. 94(1-2).

Sheskin DJ. 2004. Handbook of parametric and nonparametric statistical procedures. Florida, USA: Chapman and Hall/CRC. 1184 p.

Shuttleworth M. 2008. Design of Experiments [Online]. Available: http://www.experiment-resources.com/design-of-experiment.html 01/08/2010.

Skaggs RW. 2007. Criteria for Calculating Drain Spacing and Depth. Transactions of the ASABE. 50(5):1657-1662.

Soares JV, Almeida AC. 2001. Modelling the water balance and soil water fluxes in a fast growing Eucalyptus plantation in Brazil. Elsevier Science Publishers B.V. 253:130-147.

Société Générale d'Aménagements Hydro-Agricoles (SOGETHA). 1963. Dossier d'appel d'offres d'aménagement d'hydraulique rurale de la plaine de Tiéfora [Tender for supply dossier for rural hydraulic developpement scheme of the Tiéfora plain]. Ouagadougou (Burkina Faso). 61 p.French.

Sokona MEB, Boro A, Hema A, Katiella B. 2010. Diagnostic of the irrigated rice scheme of Tiéfora, province of Comoe, region of the Cascades. Ouagadougou (Burkina Faso): International Institute for Water and Environmental Engineering (2iE). 67 p.French.

Somado EA, Guei RG, Keya SO (eds.). 2008. NERICA: The New Rice for Africa - A Compendium. Benin: Africa Rice Center. 210 p.

Spaargaren OC, Deckers JA. 2004. Climate. In: Encyclopedia of soils in the environment. Hillel D, Rosenzweig C, Powlson D, Scow K, Singer M & Sparks D (eds.). New York, USA: Academic Press. p. 512-519.

Spiegel M, Schiller J, Srinivasan A. 2001. Probability and statistics. New York, USA: McGraw-Hill. 159 p.

Spoor G, Leeds-Harrison PB, Godwin RJ. 1982. Potential role of soil density and clay mineralogy in assessing the suitability of soils for mole drainage. Journal of Soil Science. 33(3):427-441.

Steiner KG. 1998. Using farmers' knowledge of soils in making research results more relevant to field practice: Experiences from Rwanda. Agriculture Ecosystems and Environment. 69:191-200.

Suryadi FX. 1996. Soil and water management strategies for tidal lowlands in Indonesia. Doctorate. Deft, The Netherlands: Delft University of Technology.

Tanji KK, Kielen NC. 2002. Management of irrigation drainage water. Vol.N°61. Rome (Italy): FAO. 205 p.

Thornthwaite CW. 1948. An approach toward a rational classification of climate. Geographical Review. 38(1):55-94.

Traore AS, Hatchikian CE, Belaich J-P, Le Gall J. 1981. Microcalorimetric studies of the growth of sulfate-reducing bacteria: energetics of Desulfovibrio Vulgaris growth. Journal of Bacteriology. 145(1):191-199.

Traore AS, Hatchikian CE, Le Gall J, Belaich J-P. 1982. Microcalorimetric studies of the growth of sulfate-reducing bacteria: comparison of the growth parameters of some Desulfovibrio species. Journal of Bacteriology. 49(2):606-611.

Tsay RS. 2005. Analysis of financial time series. 2nd ed. New Jersey, USA: John Wiley & Sons.

Tukey JW. 1977. Exploratory Data Analysis. Philippines: Addison-Wesley Publishing Company. 705 p.

United Nations Food and Agriculture Organization (FAO). 1993. An explanatory note on the FAO world soil resources map at 1:25 000 000 scale, 1991. Rev 1993. Rome (Italy): FAO.

United Nations Food and Agriculture Organization (FAO). 1996. Trends on Yield and Productivity of Modern Rice in Irrigated Rice Systems in Asia. In: [FAO] FaAO, ed.The FAO Expert Consultation on Technological Evolution and

Impact for Sustainable Rice Production in Asia and the Pacific; 1996, Thailand. Food and Agriculture Organisation [FAO].

United Nations Food and Agriculture Organization (FAO), INERA, SP/CPSA, PSSA, PAFR, PDA, UNC-CBF, PRP. 2006. Manuel technique d'aménagement des bas-fonds [Technical handbook for valley bottoms development]. Ouagadougou (Burkina Faso): Minitère de l'Agriculture de l'Hydraulique et des Ressources Halieutiques. 84 p.French.

United States Department of Agriculture (USDA). 1997. FM-5-410 - Millitary soils engineering. Washington, USA: US Department of the Army. 462 p.

Varani G-F, Meikle WPS, Spyromilio J, Allen DA. 1990. Direct observation of radioactive decay in Supernova 1987A. Monthly Notices of the Royal Astronomical Society. 245(1):570-576.

Vieux BE, Farajalla NS. 1994. Capturing the essential spatial variability in distributed hydrological modelling: Hydraulic roughness. Hydrological processes. 8(3):221-236.

Wailes EJ. 2005. Rice: Global trade, protectionist policies, and the impact of trade liberalization. In: Global Agrulural trade and developing countries. Ataman M & Beghin JC (eds.). Washington, D.C.: The Wolrd Bank. p. 327.

Watanabe I, Furusaka C. 1980. Microbial ecology of flooded rice soils. Advances in microbial ecology. 4:125-168.

Welch BL. 1947. The generalization of Student's Problem when several different population variances are involved. Biometrika. 34(1-2):28-35.

Wenjing L, Hongtao W, Changyong H, Reichardt W. 2008. Aromatic compound degradation by iron reducing bacteria isolated from irrigated tropical paddy soils. Journal of Envrionmental Sciences. 20(12):1487-1493.

Wesström I, Messing I, Linnér H, Lindström J. 2001. Controlled drainage - effects on drain outflowand water quality. Agricultural Water Management. 47:85-100.

West Africa Rice Development Association (WARDA). 2002. Painting the rice red: iron toxicity in the lowlands. Bouaké (Côte d'Ivoire). 9 p.

West Africa Rice Development Association (WARDA). 2006. Iron toxicity in rice-based systems in West Africa. Cotonou (Benin). 175 p.

Wilcock AA. 1968. Köppen after fifty years. Annals of the Association of American Geographers. 58:12-28.

Wind T, Conrad R. 1995. Sulfur compounds potential turnover of sulfate and thisulfate, and numbers of sulfate-reducing bacteria in planted and unplanted paddy soil. FEMS Microbiology. Ecology. 18:257-266.

Wopereis M, Donovan C, Nebié B, Guindo D, N'Diaye MK. 1999. Soil fertility management in irrigated rice systems in West Africa Part I. Agronomic analysis. Elsevier Science Publishers B.V. 61:125-145.

Yoshida T, Paul EA, MacLanen. 1976. Microbial metabolism of flooded rice soils. Soil Biochemistry. 3:83-122.

Zepenfeldt TE, Vlaar JCJ. 1990. Mise en valeur des bas-fonds en Afrique de l'Ouest. Synthèse préliminaire de l'état des connaissances [Development of valley bottoms in West Africa. Preliminary synthesis and knowledge statute]. Ouagadougou (Burkina Faso): CIEH. 137 p.French.

ANNEXES

A. List of symbols

Symbol	Description
%Clay	Percentage of clay
%OM	Percentage of organic matter
ACF	Autocorrelation Function
Af	Tropical Rainforest
Al	Aluminium
Am	Tropical Monsoon
AR	Auto Regressive
Aw	Tropical Savannah
Ca	Calcium
CEC	Cation Exchange Capacity
Db	Dry bulk density
DF	Degree of Freedom
DO	Dissolved oxygen concentration
DR	Downstream Region
Drng-	No drainage
Drng+	Drainage of 10 mm/day
DWU	Daukt water use
ETo	Reference evapo transpiration
F	Fisher statistic
Fe^{2+}	Ferrous iron concentration
Hmeas	Water depth measured with a diver
It	Infiltration rate
K	Potassium
LB	Left Bank
Lime-	No lime incorporation
Lime+	lime incorporated at the rate of 1 kg/m² of soil surface for each $\Delta pH=+1$
MA	Moving average
MAD	Mean Absolute Deviation
MAPE	Mean Absolute Percentage Error
Mg	Magnesium
MR	Middle Region
MS	Mean squares
MSD	Mean Squared Deviation
N	Nitrogen
Na	Sodium
O	Oxygen
-OH	Hydroxil group
ORP	Oxido reduction potential

Symbol	Description
P	Phosphorus
PACF	Partial Autocorrelation Function
pH	Cologarithme of proton H+ concentration
Pix	Piezometer x
p-value	Probality of obtaining the observed statistic if the null hypothesis Ho was true
RB	Right Bank
SCU	Sulphur parameter
Si	Silicium
SS	Sum of squares
STU	Sulphide Time Units
ToxScore	IRRI iron toxicity score
UR	Upstream Region
VA	Valley Axis
α	Significance level for testing p-value

B. Acronyms

Acronym	Description
ANOVA	Analysis Of Variance
ANOVA GLM	Analysis of Variance General Linear Model
ARIMA	Autoregressive Integrated Moving Averale
ASTM	American Society for Testing and Materials
FAO	United Nations Food and Agriculture Organization
IN.ERA	Institut de l'Environnement et de Recherches Agricoles
INSD	Institut National de la Statistic et de la Demographie (National Sensus Institute)
IRB	Iron Reducing Bacteria
IRRI	Internationa Rice Research Institute
MAHRH	Ministère de l'Agriculture, de l'Hydraulique et des Ressources Halieutiques (Ministry of Agriculture, Water and Fisheries Resources)
NERICA	New Rice for Africa
ONBAH	Office National des Barrages et des Aménagement Hydroagricoles
SRB	Sulphate Reducing Bacteria
UAT	Unité d'Appui Technique (Technical Assistance Unit)
USDA	United States Department of Agriculture
WARDA	West Africa Rice Development Association
ZAT	Zone d'Appui Technique (Technical Assistance Zone)

C. Bucket experiment data tables

Data Table 1: Bucket experiment *pH* data and *ORP* data

Run order	Date(days)	Drng_Id	Lime_Id	Bucket_Id	pH	ORP(mv)
1	1	Drng-	Lime+	6	6.7	26.00
2	1	Drng-	Lime+	6	6.9	19.50
3	1	Drng+	Lime+	7D	7.0	8.30
4	1	Drng-	Lime+	8	7.2	-0.40
5	10	Drng+	Lime-	1D	6.1	55.60
6	10	Drng-	Lime-	2	3.9	182.50
7	10	Drng+	Lime-	3D	6.8	15.20
8	10	Drng-	Lime-	4	4.1	171.30
9	10	Drng+	Lime+	5D	7.1	-2.90
10	10	Drng-	Lime+	6	6.9	9.30
11	10	Drng+	Lime+	7D	7.1	-2.00
12	13	Drng+	Lime-	1D	6.0	64.70
13	13	Drng-	Lime-	2	3.8	189.60
14	13	Drng+	Lime-	3D	6.2	49.90
15	13	Drng-	Lime-	4	4.2	163.50
16	13	Drng+	Lime+	5D	7.6	-29.00
17	13	Drng-	Lime+	6	7.6	-27.10
18	13	Drng+	Lime+	7D	7.3	-9.60
19	13	Drng-	Lime+	8	7.4	-17.50
20	14	Drng+	Lime-	1D	6.3	46.70
21	14	Drng-	Lime-	2	4.0	183.00
22	14	Drng-	Lime-	4	3.6	201.50
23	14	Drng+	Lime+	5D	7.4	-21.00
24	14	Drng-	Lime+	6	7.1	-1.60
25	14	Drng+	Lime+	7D	7.3	-10.20
26	15	Drng-	Lime-	2	3.9	185.20
27	15	Drng+	Lime-	3D	6.2	51.90
28	15	Drng-	Lime-	4	3.8	187.20
29	15	Drng+	Lime+	5D	7.4	-19.30
30	15	Drng-	Lime+	6	7.2	-4.90
31	15	Drng+	Lime+	7D	7.6	-28.60
32	15	Drng-	Lime+	8	7.0	4.90
33	16	Drng-	Lime-	2	3.8	196.50
34	16	Drng+	Lime-	3D	6.5	34.90

Run order	Date(days)	Drng_Id	Lime_Id	Bucket_Id	pH	ORP(mv)
35	16	Drng-	Lime-	4	3.8	192.20
36	16	Drng+	Lime+	5D	8.0	-53.90
37	16	Drng-	Lime+	6	7.1	0.60
38	16	Drng+	Lime+	7D	8.4	-79.90
39	17	Drng+	Lime-	1D	6.4	39.80
40	17	Drng-	Lime-	2	3.9	185.40
41	17	Drng+	Lime-	3D	6.6	30.50
42	17	Drng-	Lime-	4	4.0	184.90
43	17	Drng+	Lime+	5D	7.5	-23.40
44	17	Drng+	Lime+	7D	7.6	-28.50
45	20	Drng+	Lime-	1D	6.5	37.20
46	20	Drng-	Lime-	2	4.3	163.50
47	20	Drng+	Lime-	3D	6.5	35.20
48	20	Drng-	Lime-	4	4.0	181.60
49	20	Drng+	Lime+	5D	7.1	1.50
50	20	Drng+	Lime+	7D	7.4	-18.50
51	21	Drng+	Lime-	1D	6.6	29.90
52	21	Drng-	Lime-	2	4.0	178.40
53	21	Drng-	Lime-	2	4.0	180.70
54	21	Drng+	Lime-	3D	6.4	37.10
55	21	Drng-	Lime-	4	3.9	184.90
56	21	Drng-	Lime-	4	4.5	151.80
57	21	Drng+	Lime+	5D	7.2	-5.10
58	21	Drng+	Lime+	5D	6.7	24.90
59	21	Drng-	Lime+	6	6.7	24.60
60	21	Drng+	Lime+	7D	7.5	-22.90
61	21	Drng+	Lime+	7D	7.0	3.90
62	22	Drng+	Lime-	1D	6.9	12.60
63	22	Drng+	Lime-	1D	5.9	70.60
64	22	Drng-	Lime-	2	4.1	174.70
65	22	Drng-	Lime-	2	4.0	176.80
66	22	Drng-	Lime-	4	3.7	200.40
67	22	Drng-	Lime-	4	3.6	202.70
68	22	Drng+	Lime+	5D	7.4	-16.00
69	22	Drng+	Lime+	5D	7.0	7.60
70	22	Drng-	Lime+	6	7.3	-11.30
71	22	Drng+	Lime+	7D	7.6	-33.20

Run order	Date(days)	Drng_Id	Lime_Id	Bucket_Id	pH	ORP(mv)
72	22	Drng-	Lime+	8	7.3	-15.50
73	27	Drng+	Lime-	1D	6.4	71.40
74	27	Drng-	Lime-	2	5.1	135.00
75	27	Drng+	Lime-	3D	7.1	31.20
76	27	Drng-	Lime-	4	4.3	179.80
77	27	Drng-	Lime-	4	4.3	178.10
78	27	Drng+	Lime+	5D	8.0	-16.80
79	27	Drng+	Lime+	5D	7.4	18.80
80	27	Drng-	Lime+	6	8.1	-21.60
81	27	Drng-	Lime+	6	7.3	24.30
82	27	Drng+	Lime+	7D	7.6	3.70
83	27	Drng-	Lime+	8	8.0	-15.30
84	27	Drng-	Lime+	8	7.3	21.30
85	27	Drng-	Lime+	8	8.4	-35.30
86	28	Drng+	Lime-	1D	6.8	47.70
87	28	Drng-	Lime-	2	4.4	173.00
88	28	Drng+	Lime-	3D	7.2	29.60
89	28	Drng-	Lime-	4	4.4	170.70
90	28	Drng+	Lime+	5D	8.1	-18.20
91	28	Drng-	Lime+	6	8.1	-19.10
92	28	Drng+	Lime+	7D	8.0	-16.80
93	30	Drng+	Lime-	1D	7.9	-10.30
94	30	Drng-	Lime-	2	6.0	89.20
95	30	Drng+	Lime-	3D	8.2	-22.00
96	30	Drng-	Lime-	4	5.6	110.00
97	30	Drng+	Lime+	5D	8.0	-14.10
98	30	Drng-	Lime+	6	8.1	-18.10
99	30	Drng+	Lime+	7D	8.3	-28.60
100	30	Drng-	Lime+	8	8.0	-15.80
101	31	Drng+	Lime-	1D	6.5	64.50
102	31	Drng-	Lime-	2	5.3	128.60
103	31	Drng+	Lime-	3D	8.0	-12.10
104	31	Drng-	Lime-	4	4.7	156.20
105	31	Drng+	Lime+	5D	8.4	-38.50
106	31	Drng-	Lime+	6	8.1	-20.70
107	31	Drng+	Lime+	7D	8.1	-23.70
108	31	Drng-	Lime+	8	7.5	9.00

Run order	Date(days)	Drng_Id	Lime_Id	Bucket_Id	pH	ORP(mv)
109	34	Drng+	Lime-	1D	6.4	70.50
110	34	Drng-	Lime-	2	4.8	153.20
111	34	Drng+	Lime-	3D	7.1	33.60
112	34	Drng+	Lime-	3D	6.5	64.60
113	34	Drng-	Lime-	4	4.4	172.80
114	34	Drng+	Lime+	5D	7.8	-1.90
115	34	Drng-	Lime+	6	7.9	-8.30
116	34	Drng+	Lime+	7D	8.3	-28.80
117	34	Drng-	Lime+	8	8.1	-18.20
118	35	Drng+	Lime-	1D	6.6	56.90
119	35	Drng-	Lime-	2	4.2	182.10
120	35	Drng+	Lime-	3D	6.5	64.20
121	35	Drng-	Lime-	4	4.3	178.10
122	35	Drng+	Lime+	5D	7.8	-6.40
123	35	Drng-	Lime+	6	7.8	-6.20
124	35	Drng+	Lime+	7D	8.2	-26.30
125	35	Drng-	Lime+	8	7.7	1.00
126	36	Drng+	Lime-	1D	6.8	48.40
127	36	Drng-	Lime-	2	4.5	165.10
128	36	Drng-	Lime-	4	4.3	175.70
129	36	Drng+	Lime+	5D	7.8	-1.60
130	36	Drng-	Lime+	6	7.9	-7.40
131	36	Drng+	Lime+	7D	8.0	-13.10
132	36	Drng-	Lime+	8	8.2	-23.50
133	37	Drng+	Lime-	1D	6.8	45.40
134	37	Drng+	Lime-	1D	8.4	-33.70
135	37	Drng-	Lime-	?	4.6	160.40
136	37	Drng+	Lime-	3D	6.6	59.70
137	37	Drng-	Lime-	4	5.5	115.80
138	37	Drng-	Lime+	6	8.4	-34.40
139	37	Drng+	Lime+	7D	8.3	-29.90
140	37	Drng-	Lime+	8	8.3	-28.40
141	38	Drng+	Lime-	1D	6.7	53.00
142	38	Drng-	Lime-	2	4.6	161.60
143	38	Drng+	Lime-	3D	7.3	24.00
144	38	Drng-	Lime-	4	4.4	168.50
145	38	Drng+	Lime+	5D	7.7	2.70

Run order	Date(days)	Drng_Id	Lime_Id	Bucket_Id	pH	ORP(mv)
146	38	Drng-	Lime+	6	8.1	-19.10
147	38	Drng+	Lime+	7D	8.2	-23.90
148	38	Drng-	Lime+	8	8.2	-22.20
149	42	Drng+	Lime-	1D	7.2	29.10
150	42	Drng-	Lime-	2	4.7	157.00
151	42	Drng+	Lime-	3D	7.2	26.20
152	42	Drng-	Lime-	4	4.6	161.40
153	42	Drng+	Lime+	5D	7.6	4.60
154	42	Drng+	Lime+	7D	8.6	-42.50
155	42	Drng-	Lime+	8	7.7	2.00
156	43	Drng+	Lime-	1D	7.1	31.20
157	43	Drng-	Lime-	2	4.5	171.20
158	43	Drng+	Lime-	3D	7.3	21.90
159	43	Drng-	Lime-	4	4.3	176.80
160	43	Drng+	Lime+	5D	7.9	-8.40
161	43	Drng+	Lime+	7D	8.2	-25.70
162	45	Drng+	Lime-	1D	7.2	26.50
163	45	Drng-	Lime-	2	4.4	166.50
164	45	Drng+	Lime-	3D	7.4	17.00
165	45	Drng-	Lime-	4	4.4	170.70
166	45	Drng+	Lime+	5D	7.9	-8.40
167	45	Drng-	Lime+	6	8.4	-33.90
168	45	Drng+	Lime+	7D	8.2	-21.10
169	45	Drng-	Lime+	8	8.0	-11.10
170	48	Drng+	Lime-	1D	7.3	20.30
171	48	Drng-	Lime-	2	5.0	143.80
172	48	Drng+	Lime-	3D	7.4	18.20
173	48	Drng-	Lime-	4	5.0	143.00
174	48	Drng+	Lime+	5D	7.7	1.20
175	48	Drng+	Lime+	7D	7.9	-9.60
176	48	Drng-	Lime+	8	7.7	2.70
177	49	Drng+	Lime-	1D	7.5	12.50
178	49	Drng-	Lime-	2	4.6	159.40
179	49	Drng+	Lime-	3D	7.5	12.50
180	49	Drng-	Lime-	4	4.6	159.20
181	49	Drng+	Lime+	5D	7.8	-1.40
182	49	Drng-	Lime+	8	7.8	-2.90

Run order	Date(days)	Drng_Id	Lime_Id	Bucket_Id	pH	ORP(mv)
183	50	Drng+	Lime-	1D	7.9	-7.10
184	50	Drng-	Lime-	2	5.1	136.90
185	50	Drng+	Lime-	3D	7.7	3.30
186	50	Drng-	Lime-	4	4.4	172.90
187	50	Drng+	Lime+	5D	7.8	-5.40
188	50	Drng-	Lime+	6	7.9	-9.40
189	52	Drng-	Lime-	2	4.8	152.50
190	52	Drng+	Lime-	3D	7.4	18.70
191	52	Drng-	Lime-	4	4.3	179.50
192	52	Drng+	Lime+	5D	7.6	4.90
193	52	Drng+	Lime+	7D	7.8	-6.50
194	52	Drng-	Lime+	8	7.9	-11.10
195	55	Drng+	Lime-	1D	7.7	1.50
196	55	Drng-	Lime-	2	4.9	142.50
197	55	Drng+	Lime-	3D	7.7	0.30
198	55	Drng-	Lime-	4	4.6	155.30
199	55	Drng+	Lime+	5D	7.9	-6.30
200	56	Drng-	Lime-	2	4.7	153.80
201	56	Drng-	Lime-	4	4.4	166.60
202	56	Drng+	Lime+	5D	7.8	-4.70
203	57	Drng-	Lime-	2	4.8	152.30
204	57	Drng+	Lime-	3D	7.5	12.50
205	57	Drng-	Lime-	4	4.6	167.80
206	57	Drng+	Lime+	5D	7.7	0.40
207	58	Drng+	Lime-	1D	7.5	9.40
208	58	Drng-	Lime-	2	4.9	147.30
209	58	Drng+	Lime-	3D	7.6	6.50
210	58	Drng-	Lime-	4	4.5	169.50
211	58	Drng+	Lime+	5D	7.6	6.90
212	59	Drng-	Lime-	2	5.0	143.90
213	59	Drng+	Lime-	3D	7.5	13.40
214	59	Drng-	Lime-	4	5.1	140.70
215	59	Drng+	Lime+	5D	7.7	1.80
216	59	Drng+	Lime+	7D	8.4	-36.80
217	59	Drng-	Lime+	8	7.7	0.10
218	64	Drng-	Lime-	2	5.4	121.30
219	64	Drng-	Lime-	4	4.9	146.50

Run order	Date(days)	Drng_Id	Lime_Id	Bucket_Id	pH	ORP(mv)
220	64	Drng+	Lime+	5D	7.7	-1.10
221	64	Drng-	Lime+	6	7.9	-10.50
222	64	Drng+	Lime+	7D	7.9	-8.90
223	65	Drng-	Lime-	2	5.5	116.70
224	65	Drng+	Lime-	3D	7.5	13.00
225	65	Drng-	Lime-	4	5.7	106.50
226	65	Drng+	Lime+	5D	7.7	0.30
227	65	Drng+	Lime+	7D	7.8	-7.60
228	65	Drng-	Lime+	8	7.7	0.40
229	69	Drng+	Lime-	1D	7.4	15.60
230	69	Drng-	Lime-	2	5.3	129.90
231	69	Drng+	Lime-	3D	7.2	26.10
232	69	Drng-	Lime-	4	5.5	117.60
233	69	Drng+	Lime+	5D	7.3	20.10
234	69	Drng-	Lime+	6	7.6	6.80
235	69	Drng-	Lime+	8	7.6	7.30
236	71	Drng-	Lime-	2	5.3	124.60
237	71	Drng+	Lime-	3D	7.4	17.80
238	71	Drng-	Lime-	4	5.9	94.80
239	71	Drng-	Lime+	6	7.8	-6.70

D.   Tiefora farmers and plot sizes

Data Table 2: Tiefora farmers, plots and ferrous iron presence

Farm ID	Farmer name	Farm size (ha)	Iron Fe^{2+}
P 01	Bakary Koné	0.47	yes
P 02	Siaka Koné	0.40	yes
P 03	Tené Koné	0.48	yes
P 04	Abdoulaye Traoré	0.46	yes
P 05	Moulaye Sagno	0.49	yes
P 06	Groupement des femmes alakabo	0.53	yes
P 07	Baba Sory	0.48	yes
P 08	Koné Bourama dit Tassa Bra	0.54	yes
P 09	Koné Maman	0.49	yes
P 10	Yacouba Koné	0.49	yes
P 11	Daouda Koné	0.48	yes
P 12	Sory Fatouma dite Yopouo	0.45	yes
P 13	Groupement des femmes Mougnou	0.44	yes
P 14	Sicla Sory	0.41	yes
P 15	Koné Drissa	0.43	yes
P 16	Brama Tou	0.52	yes
P 17	Sali Koné	0.51	yes
P 18	Massarata Sory	0.49	yes
P 19	Maïmouna Youale Sagno	0.53	yes
P 20	Karidja Male Sory	0.47	yes
P 21	Toroba Coulibaly	0.44	yes
P 22	Nangouandji Sory	0.36	yes
P 23	Ex Sanfouain	0.32	
P 24	Lassina coulibaly	0.42	yes
P 25	Sibiri Koné président	0.51	yes
P 26	Massiatta Ouattara	0.25	yes
P 27	Ex Bénoit Somé	0.25	yes
P 28	Guino Lallé SORY	0.28	yes
P 29	Ex Téné Sory	0.23	
P 30	Gaoussou Koné	0.22	yes
P 31	Lamine Sory	0.25	yes
P32	Augustin Sory	0.52	yes
P 33	Koné Adama	0.52	yes

Farm ID	Farmer name	Farm size (ha)	Iron Fe^{2+}
P 34	Téné Sory	0.24	yes
P 35	N'galifin	0.25	yes
P 36	Bamba bassanassi dit Kouassi	0.22	yes
P 37	Minata Sory Vice pré	0.29	no
P 38	Aboubakar' Ouattara	0.47	yes
P 39	Yacouba Sory	0.23	yes
	Average	0.41	
	StDev	0.11	
	Size	39	
	Conf Int	0.03	
	Max	0.54	
	Min	0.22	
	Total	15.83	

E. Moussodougou farmers and plot sizes

Data Table 3: Moussodougou farmers and plots

Farm ID	Farmer name	Farm size (ha)	Iron Fe^{2+}
1	Hebie Baba	0.09	no
2	Hebie Soumaila	0.26	yes
3	Sourabie Abou	0.21	no
4	Sourabie Ladji	0.08	no
5	Hebie T Samba	0.20	no
6	Hebie Seydou	0.20	no
7	Diarra Baba	0.24	no
8	Sombie Gouhimane	0.22	yes
9	Hebie Dimakra	0.22	yes
10	Karama Baba	0.25	yes
11	Sourabie Daouda	0.24	yes
12	Sourabie Tenin	0.25	yes
13	Sourabie Assetou	0.22	yes
14	Tounibi Moussa	0.23	no
15	Sourabie Neyoune	0.27	no
16	Souratie C Tie	0.28	yes
17	Hetie Sapie	0.24	no
18	Siritie Djeneba	0.23	no
19	Hebie Y Drissa	0.23	no
20	Souratie Koubace	0.23	no
21	Souratie Bamogobace	0.12	yes
22	Sourabie Bassouleymane	0.22	no
23	Souratie Sita	0.21	no
24	Souratie Amadine	0.21	no
25	Sontie Mougrobahala	0.22	no
26	Sourabie S Toura	0.23	yes
27	Sourabie M Gaoussou	0.16	yes
28	Souratie Tiakolama	0.19	yes
29	Hebie Siaka	0.24	yes
30		0.15	yes
31	Souratie Tiakolama	0.11	yes
32	Souratie M Gaoussou	0.18	yes
33	Traore Moussa	0.23	yes
34		0.14	yes

Farm ID	Farmer name	Farm size (ha)	Iron Fe^{2+}
35	Souratie Fatoumata	0.14	yes
36	Souratie Mariam	0.14	yes
37	Mr Le Prefet	0.24	yes
38	Sontie Mougrobahala	0.09	yes
39		0.12	yes
40	Souratie Binta	0.24	yes
41		0.14	yes
42	Souratie Aby	0.16	yes
43	Sombie Salif	0.24	yes
44	Siritie Sali	0.23	yes
45	Wari Tiecoura	0.24	yes
46	Sombie C Bie	0.28	yes
47	Souratie Aby	0.31	yes
48	Sourabie Batchemne	0.23	yes
49	Hebie Abou	0.17	yes
50	Sombie Panhol	0.17	yes
51	Sourabie Tianbeko	0.24	yes
52	Hebie Bangale M	0.18	yes
53	Ouattara Tiemoko	0.18	yes
54	Hebie Abou	0.14	yes
55	Karama Baba	0.22	yes
56	Souratie Bintou	0.24	yes
57	Souratie Djata	0.23	yes
58	Sourabie Tienhale	0.23	yes
59	Souratie Adama	0.24	yes
60		0.25	yes
61	Hebie Mamadou	0.22	yes
62	Hebie Baba	0.11	yes
63	Sourabie Lacine	0.18	yes
64	Hebie Bie	0.24	yes
65	Sourabie Ladji	0.25	yes
66		0.28	yes
67	Sourabie Drissa	0.23	yes
68	Sombie Panhol	0.11	yes
69	Sourabie Bahilama	0.18	yes
70	Sourabie Mounihaly	0.18	yes
71	Siribie Amadou	0.16	yes

Farm ID	Farmer name	Farm size (ha)	Iron Fe^{2+}
72	Sombie Moussa N'dyane	0.20	yes
73	Souratie Norhela	0.22	yes
74	Hebie Seydou	0.10	yes
75	Souratie Tiebateli	0.08	yes
76	Sourabie Oumar	0.10	yes
77		0.08	yes
78	Sourabie Foromine	0.22	yes
79	Sourabie Hlement	0.25	yes
80	Nedonbie Sidiki	0.23	yes
81		0.20	yes
82	Souratie Tenin Et Sourabie Nouyoune B	0.24	yes
83	Souratie Tenin Et Sourabie Nouyoune B	0.18	yes
84	Souratie Tenin Et Sourabie Nouyoune B	0.24	yes
85		0.30	yes
86		0.12	yes
87		0.48	yes
	Quatier Dou	0.00	
88	Sombie Yacouba	0.18	yes
89	Amadou Diarra	0.17	yes
90	Karama Baba	0.25	yes
91	Sombie Sekou	0.24	yes
92	Hebie Seydou	0.22	yes
93		0.25	yes
94		0.24	yes
95	Sourabie Djela	0.12	yes
96	Sombie Yacouba	0.30	yes
97	Mr Le Maire	0.24	yes
98	Kadidjadal Sontie	0.23	yes
99	Sourabie Bessala	0.23	yes
100	Hebie Siaka	0.23	yes
101	Sontie Aminata	0.13	yes
102	Hebie Lacine	0.23	yes
103	Hebie Bramah	0.24	yes
104	Sombie Drissa	0.23	yes
105	Sombie Aly	0.22	yes
106	Sombie Souleymane	0.12	yes
107	Sourabie Souleymane	0.24	yes

Farm ID	Farmer name	Farm size (ha)	Iron Fe^{2+}
108	Sourabie Djibril	0.24	yes
109	Sourabie Noufoun	0.26	yes
110	Souratie Barke	0.15	yes
111	Sourabie Amara	0.20	yes
112	Traore Tchefi	0.27	yes
113	Sombie Massa	0.15	yes
114	Sombie Massa	0.09	yes
115	Diallo Mamoudou	0.14	yes
116	Sontie Djenebou	0.17	yes
117	Hebie Goyoroba	0.15	yes
118	Souratie Awani	0.25	yes
119	Sombie Yaledjon	0.06	yes
120	Sourabie Neyoune S	0.21	yes
121		0.20	yes
122	Barry Razogo	0.23	yes
123	Sombie Sibiri	0.24	yes
124	Hebie Allakabo	0.26	yes
125	Hebie Modo	0.16	yes
126		0.21	yes
127	Sombie Tinibie Moussa	0.28	yes
128	Sourabie Kanawali	0.28	yes
129		0.26	yes
130	Sombie Dimakro	0.16	yes
131	Sourabie Daouda	0.26	yes
132	Hebie Tchalemon	0.28	yes
133		0.13	yes
134	Sombie Drissa	0.24	yes
135	Souratie Kadidja	0.22	yes
136	Sourabie Kayal L	0.13	yes
137	Sourabie Djelila	0.18	yes
138	Sourabie Djelila	0.19	yes
139	Sontie Tenin	0.21	yes
140	Sourabie Noumbie	0.23	yes
141	Hebie Adama	0.23	yes
142	Sombie Baba	0.21	yes
143	Sourabie Teli	0.22	yes
144	Sourabie Sadjagousse	0.25	yes

Farm ID	Farmer name	Farm size (ha)	Iron Fe^{2+}
145	Sourabie Goutocce	0.22	yes
146	Sontie Adamani	0.20	yes
147	Sombie Yalmon	0.15	yes
148	Souratie Sita	0.23	yes
149	Sombie Yacouba	0.24	yes
150	Sourabie Madjambi	0.17	yes
151	Gpr Massinimissi	0.22	yes
152	Sourabie Gbanbra	0.22	yes
153	Gpr Massinimissi	0.26	yes
154	Ouattara Tiemoko	0.24	yes
155	Sourabie Mounihaly	0.19	yes
156	Sourabie Noubie	0.31	yes
157	Sourabie Teli	0.24	yes
158	Hebie Adama	0.22	yes
159	Sombie Massa	0.24	yes
160	Souratie Koubace	0.23	yes
161	Sourabie Yougoubi	0.23	yes
162	Sourabie Noubie	0.24	yes
163	Sourabie Bassouleymane	0.25	yes
164	Gpr Massinimissi	0.20	yes
165	Sombie Yalmon	0.25	yes
166	Ousmane Sourabie	0.23	yes
167	Sourabie Goyoroba	0.24	yes
168	Sourabie Namartchalil	0.25	yes
169	Sourabie Lamine	0.13	yes
	Size	169	
	Average	0.21	
	Stdev	0.06	
	Confidence Interval	0.01	
	Total	35.21	

F.   Microplot experiments data and results

Data Table 4: Microplot experiment *pH*, *DO* and *Fe²⁺* data

Run order	Drng_ID	Depth_Id	Date (days)	Meas_ID	pH	DO (mg/l)	Fe²⁺(mg/l)
1	Drng-	Depth-	1	P1.1	5.3	1.88	530
2	Drng-	Depth-	1	P1.1	5.8	3.27	420
3	Drng-	Depth-	1	P1.1	4.4	0.78	750
4	Drng-	Depth-	1	P1.3	4.8	0.18	570
5	Drng-	Depth-	1	P1.3	4.6	0.18	730
6	Drng-	Depth-	1	P1.3	4.8	0.12	780
7	Drng-	Depth-	1	P2.2	5.3	0.35	1270
8	Drng-	Depth-	1	P2.2	5.3	6.59	910
9	Drng-	Depth-	1	P2.2	4.8	0.25	1050
10	Drng-	Depth-	1	P3.1	4.9	0.16	270
11	Drng-	Depth-	1	P3.1	4.5	0.29	750
12	Drng-	Depth-	1	P3.1	4.5	0.08	940
13	Drng-	Depth-	1	P3.3	4.8	0.33	540
14	Drng-	Depth-	1	P3.3	4.8	0.17	500
15	Drng-	Depth-	1	P3.3	4.7	0.11	570
16	Drng-	Depth-	1	P4.2	4.7	0.12	930
17	Drng-	Depth-	1	P4.2	5.5	0.17	330
18	Drng-	Depth-	1	P4.2	4.6	0.19	910
19	Drng+	Depth-	1	P1.2	4.6	0.54	910
20	Drng+	Depth-	1	P1.2	4.8	0.15	630
21	Drng+	Depth-	1	P1.2	4.9	2.45	590
22	Drng+	Depth-	1	P2.1	4.7	0.75	310
23	Drng+	Depth-	1	P2.1	4.0	0.29	290
24	Drng+	Depth-	1	P2.1	4.6	0.18	630
25	Drng+	Depth-	1	P2.3	5.1	0.67	610
26	Drng+	Depth-	1	P2.3	4.9	4.02	610
27	Drng+	Depth-	1	P2.3	4.5	0.62	520
28	Drng+	Depth-	1	P3.2	4.5	0.15	950
29	Drng+	Depth-	1	P3.2	4.9	0.24	530
30	Drng+	Depth-	1	P3.2	4.7	0.12	570
31	Drng+	Depth-	1	P4.1	6.9	0.31	1160
32	Drng+	Depth-	1	P4.1	4.8	3.58	1070
33	Drng+	Depth-	1	P4.1	4.6	2.68	540

Run order	Drng_ID	Depth_Id	Date (days)	Meas_ID	pH	DO (mg/l)	Fe^{2+}(mg/l)
34	Drng+	Depth-	1	P4.3	4.8	8.01	860
35	Drng+	Depth-	1	P4.3	4.3	0.22	400
36	Drng+	Depth-	1	P4.3	4.8	0.11	630
37	Drng-	Depth+	1	P1.1	4.8	0.22	930
38	Drng-	Depth+	1	P1.1	4.8	0.21	520
39	Drng-	Depth+	1	P1.1	4.5	4.27	920
40	Drng-	Depth+	1	P1.3	4.6	0.46	710
41	Drng-	Depth+	1	P1.3	4.7	0.14	550
42	Drng-	Depth+	1	P1.3	4.7	0.10	1270
43	Drng-	Depth+	1	P2.2	3.9	1.19	740
44	Drng-	Depth+	1	P2.2	4.1	0.14	650
45	Drng-	Depth+	1	P2.2	4.2	0.09	430
46	Drng-	Depth+	1	P3.1	4.5	0.07	1150
47	Drng-	Depth+	1	P3.1	5.3	0.19	1040
49	Drng-	Depth+	1	P3.3	4.4	0.09	790
50	Drng-	Depth+	1	P3.3	4.1	0.32	750
51	Drng-	Depth+	1	P3.3	4.4	0.07	800
52	Drng-	Depth+	1	P4.2	4.7	0.07	810
53	Drng-	Depth+	1	P4.2	4.8	0.07	440
54	Drng-	Depth+	1	P4.2	4.7	0.06	570
55	Drng+	Depth+	1	P1.2	4.3	0.72	550
56	Drng+	Depth+	1	P1.2	4.3	0.49	230
57	Drng+	Depth+	1	P1.2	4.1	0.67	980
59	Drng+	Depth+	1	P2.1	4.9	0.56	1260
61	Drng+	Depth+	1	P2.3	4.2	0.10	720
62	Drng+	Depth+	1	P2.3	4.7	0.41	470
63	Drng+	Depth+	1	P2.3	4.4	0.21	630
64	Drng+	Depth+	1	P3.2	4.4	0.20	550
65	Drng+	Depth+	1	P3.2	4.8	0.10	400
66	Drng+	Depth+	1	P3.2	4.1	0.24	560
67	Drng+	Depth+	1	P4.1	4.8	0.11	870
68	Drng+	Depth+	1	P4.1	4.5	0.13	760
69	Drng+	Depth+	1	P4.1	4.7	0.15	500
70	Drng+	Depth+	1	P4.3	4.5	0.39	480
71	Drng+	Depth+	1	P4.3	4.5	0.53	1110
72	Drng+	Depth+	1	P4.3	4.7	3.04	920
73	Drng-	Depth-	23	P1.1	5.4	0.39	1250

Run order	Drng_ID	Depth_Id	Date (days)	Meas_ID	pH	DO (mg/l)	Fe^{2+}(mg/l)
74	Drng-	Depth-	23	P1.1	5.1	0.19	1110
75	Drng-	Depth-	23	P1.1	5.1	0.11	680
76	Drng-	Depth-	23	P1.3	5.7	0.06	900
77	Drng-	Depth-	23	P1.3	5.2	0.15	1440
78	Drng-	Depth-	23	P1.3	5.2	0.11	1300
79	Drng-	Depth-	23	P2.2	5.0	0.06	1010
80	Drng-	Depth-	23	P2.2	5.1	0.06	730
81	Drng-	Depth-	23	P2.2	5.6	0.06	600
82	Drng-	Depth-	23	P3.1	5.2	0.10	1930
83	Drng-	Depth-	23	P3.1	5.5	0.06	600
84	Drng-	Depth-	23	P3.1	5.4	0.10	880
85	Drng-	Depth-	23	P3.3	5.0	0.08	1510
86	Drng-	Depth-	23	P3.3	5.6	0.08	550
87	Drng-	Depth-	23	P3.3	5.7	0.07	560
88	Drng-	Depth-	23	P4.2	5.2	0.06	700
89	Drng-	Depth-	23	P4.2	5.3	0.07	1230
90	Drng-	Depth-	23	P4.2	5.4	0.11	920
91	Drng+	Depth-	23	P1.2	5.3	0.43	1270
92	Drng+	Depth-	23	P1.2	6.0	0.13	1670
93	Drng+	Depth-	23	P1.2	5.7	0.12	1240
94	Drng+	Depth-	23	P2.1	4.9	0.06	540
95	Drng+	Depth-	23	P2.1	4.9	0.24	860
96	Drng+	Depth-	23	P2.1	4.9	0.14	680
97	Drng+	Depth-	23	P2.3	4.9	0.08	2180
98	Drng+	Depth-	23	P2.3	5.2	0.10	1620
99	Drng+	Depth-	23	P2.3	4.9	0.13	1729
100	Drng+	Depth-	23	P3.2	5.4	0.24	940
101	Drng+	Depth-	23	P3.2	5.3	0.09	2160
102	Drng+	Depth-	23	P3.2	5.3	0.06	720
103	Drng+	Depth-	23	P4.1	5.5	0.13	700
104	Drng+	Depth-	23	P4.1	6.2	0.08	470
105	Drng+	Depth-	23	P4.1	6.2	0.12	1230
106	Drng+	Depth-	23	P4.3	5.1	0.19	970
107	Drng+	Depth-	23	P4.3	5.0	0.08	1910
108	Drng+	Depth-	23	P4.3	4.8	0.21	1250
109	Drng-	Depth+	23	P1.1	5.8	0.15	690
110	Drng-	Depth+	23	P1.1	5.5	0.17	1340

Run order	Drng_ID	Depth_Id	Date (days)	Mcas_ID	pH	DO (mg/l)	Fe^{2+}(mg/l)
111	Drng-	Depth+	23	P1.1	5.6	0.09	1120
112	Drng-	Depth+	23	P1.3	5.5	0.06	330
113	Drng-	Depth+	23	P1.3	5.6	0.08	1120
114	Drng-	Depth+	23	P1.3	5.5	0.06	860
115	Drng-	Depth+	23	P2.2	5.4	0.02	640
116	Drng-	Depth+	23	P2.2	4.8	0.04	670
117	Drng-	Depth+	23	P2.2	4.9	0.05	1090
118	Drng-	Depth+	23	P3.1	5.1	0.06	1210
119	Drng-	Depth+	23	P3.1	4.8	0.06	680
120	Drng-	Depth+	23	P3.1	4.8	0.05	700
121	Drng-	Depth+	23	P3.3	5.7	0.07	1430
122	Drng-	Depth+	23	P3.3	5.2	0.05	880
123	Drng-	Depth+	23	P3.3	5.0	0.06	870
124	Drng-	Depth+	23	P4.2	5.8	0.03	1230
125	Drng-	Depth+	23	P4.2	5.5	0.06	1200
126	Drng-	Depth+	23	P4.2	5.3	0.05	420
127	Drng+	Depth+	23	P1.2	4.9	0.10	1550
128	Drng+	Depth+	23	P1.2	5.0	0.08	1220
129	Drng+	Depth+	23	P1.2	5.5	0.10	650
130	Drng+	Depth+	23	P2.1	5.1	0.08	1640
131	Drng+	Depth+	23	P2.1	4.8	0.10	970
132	Drng+	Depth+	23	P2.1	5.1	0.13	890
133	Drng+	Depth+	23	P2.3	5.0	0.06	1390
134	Drng+	Depth+	23	P2.3	5.1	0.06	570
135	Drng+	Depth+	23	P2.3	5.6	0.08	630
136	Drng+	Depth+	23	P3.2	4.9	0.10	860
137	Drng+	Depth+	23	P3.2	5.2	0.07	800
138	Drng+	Depth+	23	P3.2	5.3	0.06	820
139	Drng+	Depth+	23	P4.1	5.4	0.06	520
140	Drng+	Depth+	23	P4.1	5.7	0.04	1270
141	Drng+	Depth+	23	P4.1	5.3	0.06	720
142	Drng+	Depth+	23	P4.3	5.4	0.04	1460
143	Drng+	Depth+	23	P4.3	5.7	0.05	1010
144	Drng+	Depth+	23	P4.3	5.1	0.06	860
145	Drng-	Depth-	58	P1.1	6.8	0.16	1180
146	Drng-	Depth-	58	P1.1	6.7	0.17	576
147	Drng-	Depth-	58	P1.1	6.6	0.15	1427

Run order	Drng_ID	Depth_Id	Date (days)	Meas_ID	pH	DO (mg/l)	Fe^{2+}(mg/l)
148	Drng-	Depth-	58	P1.3	6.6	0.11	1020
149	Drng-	Depth-	58	P1.3	6.8	0.08	1176
150	Drng-	Depth-	58	P1.3	6.6	0.06	990
151	Drng-	Depth-	58	P2.2	6.9	0.07	1153
153	Drng-	Depth-	58	P2.2	6.5	0.07	274
154	Drng-	Depth-	58	P3.1	6.7	0.13	2525
155	Drng-	Depth-	58	P3.1	6.6	0.08	1263
156	Drng-	Depth-	58	P3.1	6.8	0.09	1043
157	Drng-	Depth-	58	P3.3	6.7	0.05	3651
158	Drng-	Depth-	58	P3.3	6.6	0.09	1180
159	Drng-	Depth-	58	P3.3	6.8	0.07	4886
160	Drng-	Depth-	58	P4.2	6.4	0.13	5435
161	Drng-	Depth-	58	P4.2	6.5	0.08	5105
162	Drng-	Depth-	58	P4.2	6.5	0.12	686
163	Drng+	Depth-	58	P1.2	6.7	0.24	710
164	Drng+	Depth-	58	P1.2	6.5	0.13	1320
165	Drng+	Depth-	58	P1.2	6.4	0.18	2230
166	Drng+	Depth-	58	P2.1	6.5	0.09	1619
167	Drng+	Depth-	58	P2.1	6.5	0.08	1812
168	Drng+	Depth-	58	P2.1	6.6	0.17	961
169	Drng+	Depth-	58	P2.3	6.9	0.17	5023
170	Drng+	Depth-	58	P2.3	7.0	0.14	2114
171	Drng+	Depth-	58	P2.3	6.6	0.13	3541
172	Drng+	Depth-	58	P3.2	6.6	0.08	4282
174	Drng+	Depth-	58	P3.2	6.9	0.14	1098
175	Drng+	Depth-	58	P4.1	6.6	0.13	7960
176	Drng+	Depth-	58	P4.1	6.7	0.10	2443
177	Drng+	Depth-	58	P4.1	6.7	0.13	906
178	Drng+	Depth-	58	P4.3	6.7	0.13	6382
179	Drng+	Depth-	58	P4.3	6.6	0.15	3678
180	Drng+	Depth-	58	P4.3	6.5	0.14	5352
181	Drng-	Depth+	58	P1.1	6.6	0.08	2306
182	Drng-	Depth+	58	P1.1	6.7	0.09	988
183	Drng-	Depth+	58	P1.1	6.6	0.09	1537
184	Drng-	Depth+	58	P1.3	6.7	0.04	1380
185	Drng-	Depth+	58	P1.3	6.6	0.04	660
186	Drng-	Depth+	58	P1.3	6.6	0.05	990

Run order	Drng_ID	Depth_Id	Date (days)	Meas_ID	pH	DO (mg/l)	Fe^{2+}(mg/l)
187	Drng-	Depth+	58	P2.2	6.5	0.05	384
188	Drng-	Depth+	58	P2.2	6.5	0.05	1345
189	Drng-	Depth+	58	P2.2	6.5	0.04	823
190	Drng-	Depth+	58	P3.1	6.9	0.09	686
191	Drng-	Depth+	58	P3.1	6.5	0.07	1894
192	Drng-	Depth+	58	P3.1	6.5	0.07	1427
193	Drng-	Depth+	58	P3.3	6.5	0.04	1235
194	Drng-	Depth+	58	P3.3	6.5	0.07	1235
195	Drng-	Depth+	58	P3.3	6.6	0.04	1263
197	Drng-	Depth+	58	P4.2	6.5	0.08	1784
198	Drng-	Depth+	58	P4.2	6.5	0.06	714
199	Drng+	Depth+	58	P1.2	6.5	0.09	930
200	Drng+	Depth+	58	P1.2	6.5	0.09	810
201	Drng+	Depth+	58	P1.2	6.4	0.14	990
202	Drng+	Depth+	58	P2.1	6.5	0.05	686
203	Drng+	Depth+	58	P2.1	6.4	0.06	933
204	Drng+	Depth+	58	P2.1	6.3	0.07	576
205	Drng+	Depth+	58	P2.3	6.5	0.05	1921
206	Drng+	Depth+	58	P2.3	6.4	0.05	357
207	Drng+	Depth+	58	P2.3	6.8	0.08	2333
208	Drng+	Depth+	58	P3.2	6.9	0.09	2525
209	Drng+	Depth+	58	P3.2	6.8	0.09	604
210	Drng+	Depth+	58	P3.2	7.0	0.05	1235
211	Drng+	Depth+	58	P4.1	6.9	0.04	439
212	Drng+	Depth+	58	P4.1	6.5	0.09	4254
213	Drng+	Depth+	58	P4.1	6.4	0.08	1757
214	Drng+	Depth+	58	P4.3	6.4	0.08	2937
215	Drng+	Depth+	58	P4.3	6.7	0.08	3637
216	Drng+	Depth+	58	P4.3	6.3	0.10	2992
217	Drng-	Depth-	86	P1.1			233
218	Drng-	Depth-	86	P1.1			
219	Drng-	Depth-	86	P1.1			947
221	Drng-	Depth-	86	P1.3	6.9	0.08	412
222	Drng-	Depth-	86	P1.3	6.8	0.11	1441
223	Drng-	Depth-	86	P2.2	6.9	1.23	618
224	Drng-	Depth-	86	P2.2	6.9	0.11	1304
225	Drng-	Depth-	86	P2.2	6.9	0.07	686

Run order	Drng_ID	Depth_Id	Date (days)	Meas_ID	pH	DO (mg/l)	Fe^{2+}(mg/l)
226	Drng-	Depth-	86	P3.1	6.6	0.08	1304
227	Drng-	Depth-	86	P3.1	6.7	0.13	892
230	Drng-	Depth-	86	P3.3	6.7	0.10	961
231	Drng-	Depth-	86	P3.3	6.5	0.12	892
232	Drng-	Depth-	86	P4.2	6.4	0.13	1304
233	Drng-	Depth-	86	P4.2	6.5	0.08	1098
236	Drng+	Depth-	86	P1.2			2539
237	Drng+	Depth-	86	P1.2			1098
238	Drng+	Depth-	86	P2.1	6.6	0.17	892
239	Drng+	Depth-	86	P2.1	6.8	0.19	686
240	Drng+	Depth-	86	P2.1	6.7	0.15	823
241	Drng+	Depth-	86	P2.3	6.6	0.13	686
242	Drng+	Depth-	86	P2.3	7.0	0.14	1167
243	Drng+	Depth-	86	P2.3	6.9	0.17	412
246	Drng+	Depth-	86	P3.2	6.6	0.08	2333
247	Drng+	Depth-	86	P4.1	6.7	0.10	1578
248	Drng+	Depth-	86	P4.1	6.6	0.13	892
249	Drng+	Depth-	86	P4.1	6.6	0.37	3431
250	Drng+	Depth-	86	P4.3	6.7	0.13	1098
253	Drng-	Depth+	86	P1.1			940
254	Drng-	Depth+	86	P1.1			364
256	Drng-	Depth+	86	P1.3			1441
258	Drng-	Depth+	86	P1.3			3225
260	Drng-	Depth+	86	P2.2	6.5	0.06	1304
262	Drng-	Depth+	86	P3.1	6.4	0.09	1098
264	Drng-	Depth+	86	P3.1	6.9	0.09	755
271	Drng+	Depth+	86	P1.2			3705
275	Drng+	Depth+	86	P2.1	6.4	0.06	2676
276	Drng+	Depth+	86	P2.1	6.3	0.07	3362
278	Drng+	Depth+	86	P2.3	6.4	0.05	1098
282	Drng+	Depth+	86	P3.2	6.9	0.09	2608
283	Drng+	Depth+	86	P4.1	6.9	0.04	1235
284	Drng+	Depth+	86	P4.1	6.5	0.09	1029
285	Drng+	Depth+	86	P4.1	6.3	0.12	961
286	Drng+	Depth+	86	P4.3	6.4	0.08	823
287	Drng+	Depth+	86	P4.3	6.7	0.08	2470
288	Drng+	Depth+	86	P4.3	6.3	0.10	3225

Data Table 5: Microplot experiment IRRI iron toxicity scores

Run order	Drng_ID	Date (days)	ToxScore
1	Drng-	01	0
2	Drng-	01	0
3	Drng-	01	0
4	Drng-	01	0
5	Drng-	01	0
6	Drng-	01	0
7	Drng-	08	0
8	Drng-	08	0
9	Drng-	08	0
10	Drng-	08	0
11	Drng-	08	0
12	Drng-	08	0
13	Drng-	15	0
14	Drng-	15	0
15	Drng-	15	0
16	Drng-	15	0
17	Drng-	15	0
18	Drng-	15	0
19	Drng-	22	0
20	Drng-	22	0
21	Drng-	22	0
22	Drng-	22	0
23	Drng-	22	0
24	Drng-	22	0
25	Drng-	29	1
26	Drng-	29	0
27	Drng-	29	1
28	Drng-	29	0
29	Drng-	29	0
30	Drng-	29	1
31	Drng-	35	1
32	Drng-	35	0
33	Drng-	35	3
34	Drng-	35	3
35	Drng-	35	0
36	Drng-	35	3
37	Drng-	43	1
38	Drng-	43	1
39	Drng-	43	3
40	Drng-	43	5

Run order	Drng_ID	Date (days)	ToxScore
41	Drng-	43	3
42	Drng-	43	5
43	Drng-	50	3
44	Drng-	50	3
45	Drng-	50	3
46	Drng-	50	5
47	Drng-	50	3
48	Drng-	50	5
49	Drng-	57	5
50	Drng-	57	5
51	Drng-	57	3
52	Drng-	57	3
53	Drng-	57	3
54	Drng-	57	5
55	Drng-	64	7
56	Drng-	64	5
57	Drng-	64	3
58	Drng-	64	5
59	Drng-	64	5
60	Drng-	64	5
61	Drng-	71	7
62	Drng-	71	7
63	Drng-	71	5
64	Drng-	71	5
65	Drng-	71	5
66	Drng-	71	7
67	Drng+	01	0
68	Drng+	01	0
69	Drng+	01	0
70	Drng+	01	0
71	Drng+	01	0
72	Drng+	01	0
73	Drng+	08	0
74	Drng+	08	0
75	Drng+	08	0
76	Drng+	08	0
77	Drng+	08	0
78	Drng+	08	0
79	Drng+	15	0
80	Drng+	15	0
81	Drng+	15	0
82	Drng+	15	0
83	Drng+	15	0

Run order	Drng_ID	Date (days)	ToxScore
84	Drng+	15	0
85	Drng+	22	0
86	Drng+	22	0
87	Drng+	22	0
88	Drng+	22	0
89	Drng+	22	0
90	Drng+	22	0
91	Drng+	29	3
92	Drng+	29	0
93	Drng+	29	1
94	Drng+	29	0
95	Drng+	29	1
96	Drng+	29	3
97	Drng+	35	3
98	Drng+	35	3
99	Drng+	35	3
100	Drng+	35	0
101	Drng+	35	1
102	Drng+	35	3
103	Drng+	43	5
104	Drng+	43	3
105	Drng+	43	5
106	Drng+	43	1
107	Drng+	43	1
108	Drng+	43	5
109	Drng+	50	3
110	Drng+	50	3
111	Drng+	50	5
112	Drng+	50	3
113	Drng+	50	3
114	Drng⏐	50	5
115	Drng+	57	5
116	Drng+	57	3
117	Drng+	57	3
118	Drng+	57	3
119	Drng+	57	5
120	Drng+	57	5
121	Drng+	64	5
122	Drng+	64	3
123	Drng+	64	5
124	Drng+	64	5
125	Drng+	64	5
126	Drng+	64	5

Run order	Drng_ID	Date (days)	ToxScore
127	Drng+	71	7
128	Drng+	71	5
129	Drng+	71	5
130	Drng+	71	5
131	Drng+	71	7
132	Drng+	71	7

G.   Samenvatting

De tegenstrijdigheid tussen de groei van de wereldbevolking en de toename van de beperkingen met betrekking tot het ontwikkelen van nieuwe irrigatie systemen vormt een bijzondere uitdaging voor de menselijke gewasproductie systemen die dient te worden genomen. De bevolking in vele landen in Azië, het Midden-Oosten en Afrika zal naar verwachting in de komende 50 jaar verdubbelen. De ervaring met de groene revolutie in Azië - waarbij 70% van de toename in de voedselproductie werd gerealiseerd door geïrrigeerde landbouw - toont aan dat het niet alleen noodzakelijk is om te streven naar een toename van dergelijke gewas productie systemen, maar ook om de productie efficiëntie van bestaande systemen te verbeteren. In feite, als een zeer verontrustende zaak, ligt de rijst productie in de geïrrigeerde dalgronden van de Afrikaanse tropische savanne ver beneden de te verwachten opbrengst. Een van de belangrijkste belemmeringen voor deze productie is ijzer toxiciteit door slechte ontwatering. Volgens Afrika Rijst, wordt ten minste 60% van de moerassige dalgronden in de tropische savanne beïnvloed door een zekere mate van ijzer toxiciteit. In veel gebieden daalt de opbrengst tot nul, met achterlating van miljoenen teleurgestelde en arme boeren. Daarom is het niet verwonderlijk dat er een sterke onderzoek dynamiek is - variërend van landbouwkunde tot microbiologie - waarbij ernaar wordt gestreefd om met oplossingen te komen voor het verminderen van ijzer toxiciteit bij rijst. Omdat overwegend zuurstofloze omstandigheden in de bodem in combinatie met de ontwikkeling van ijzer reducerende bacteriën essentiële factoren bleken te zijn voor de ontwikkeling van ijzer toxiciteit, is dit onderzoek gericht geweest op mogelijke bijdragen van ondergrondse drainage aan het oplossen van dit probleem.

Twee complementaire reeksen activiteiten - ontworpen binnen twee project componenten en gericht op vijf fundamentele vragen die nauw verwant waren aan de factoren die bijdragen aan de ontwikkeling van ijzer toxiciteit - zijn uitgevoerd. In feite werd het onderzoek uitgevoerd in twee belangrijke componenten: veldonderzoeken en ontworpen experimenten. Bij de veldonderzoeken zijn de factoren onderzocht die ijzer toxiciteit teweeg brengen of verergeren, zoals gehalten aan klei, concentratie aan ferro-ionen Fe^{2+}, opgeloste zuurstof, zuurgraad van de bodem of het waterbeheer. Profijt trekkend van de kennis die is opgedaan bij het veldonderzoek en literatuurstudie, zijn twee parallelle experimenten ontworpen met behulp van beton microplots enerzijds en emmers anderzijds, om op statistische wijze het effect van de ondergrond op de zuurgraad en de veranderingen in de ferro-ijzer concentratie te bepalen. Alle activiteiten die zijn uitgevoerd binnen de twee componenten van dit onderzoek waren erop gericht om de volgende vijf onderzoeksvragen te beantwoorden:

1) hoe verandert ferro-ijzer, dat gevormd en verspreid is in de bodem, in ijzer toxiciteit?
2) hoe is de klei verspreid in de dalgrond?
3) hoe wordt de doorlatendheid beïnvloed door de verdeling van de klei in de dalgrond?
4) hoe kan waterbeheer helpen bij het verbeteren van de bodemgesteldheid?
5) Wat is het effect van ondergrondse drainage op ijzer toxiciteit?

De antwoorden op deze onderzoeksvragen - reeds gepubliceerd of in druk - worden hieronder weergegeven, gevolgd door de bijdrage van dit onderzoek in twee gebieden: i) wetenschap en techniek, en ii) sociale economie.

Klei en ferro-ijzer kan in lagen worden afgezet

Hoge ferro ion Fe^{2+} concentratie, ingekapseld in dichte kleilagen, vormt een belangrijke bedreiging voor de rijstproductie bij verschillende irrigatie systemen in dalgronden in de tropische savanne van West-Afrika. Veel activiteiten worden momenteel ondernomen om de ijzer toxiciteit te verminderen. In deze studie hebben we de aanwezigheid van klei en ferro-ijzer gelaagdheid in een typische overstroombare dalgrond genaamd Tiefora in Burkina Faso onderzocht. Rekening houdend met de verschillende hellingen in de vallei zijn twee willekeurige monsters van de bodem op verschillende diepten genomen. De monsters zijn genomen tot een diepte van 500 cm, maar in het bijzonder op 30, 50 en 100 cm. Het klei percentage is bepaald met korrel grootte analyse. Ferro-ijzer concentraties zijn bepaald met de reflectometrische methode. De gelaagdheden van klei en ferro-ionen Fe^{2+} werden bepaald met behulp van statistische hypothese testen (ANOVA en Welch t-test). Het klei percentage over de eerste 100 cm bovengrond - 28,9% - bleek twee keer zo hoog te zijn als in de onderliggende lagen. Voorts werd ferro-ijzer voornamelijk aangetroffen in de bovenste 30 cm, met een gemiddelde concentratie van 994 mg/l. Deze ferro-ijzer concentratie is veel hoger dan op diepten van 50 en 100 cm daaronder (73 mg/l), terwijl de *pH* van de drie lagen vrijwel neutraal is. Deze opvallende gelaagdheid suggereert verschillende mogelijkheden om de ijzer toxiciteit te verminderen. Onder deze mogelijkheden, stellen wij voor het handhaven van natte omstandigheden tijdens de groeiperiode in de geïrrigeerde gronden in combinatie met uitloging door ondergrondse drainage in de perioden dat de landen braak liggen.

IJzer toxiciteit risico is hoger bij irrigatie systemen voor een enkel seizoen

Met als doel het vinden van de geochemische verschillen en te helpen bij het ontwikkelen van strategieën tegen ijzer toxiciteit, zijn twee geïrrigeerde rijst gebieden met hematiet dominante dalgronden onderzocht in de tropische savanne regio van Burkina Faso. Het eerste gebied was Tiefora, een 16 ha modern dubbel seizoen geïrrigeerd rijst systeem en matig beïnvloed door ijzer toxiciteit (10% van het gebied met een toxiciteit score van 4). Het tweede gebied was Moussodougou, een 35 ha traditioneel enkel seizoen geïrrigeerde rijst dalgrond, waarvan 50% meer ernstige ijzer toxiciteit vertoont (score 7). Negen bodem monsters zijn genomen op drie diepte - 30, 50 en 100 cm - waarvan 27 in Tiefora en 27 in Moussodougou. Vijf technieken zijn gebruikt om de gegevens te meten: i) de ferro-ijzer concentratie werd bepaald met een reflectometer; ii) een *pH* meter leverde de *pH*; iii) klei gehalten zijn verkregen met een korrel grootte analyse en dichtheidsmeter van het United States Department of Army (USDA); iv) de organische stof werd bepaald door drogen in een oven; v) de droge dichtheid werd bepaald met behulp van ongestoorde bodemmonsters. Statistische hypothese testen, eenzijdige ANOVA en Welch t-test, zijn toegepast op de gegevens om de overeenkomsten en verschillen tussen de twee locaties vast te leggen. Een geochemische analyse volgde om de oorzaken van deze verschillen te vinden. De resultaten toonden aan dat terwijl de oxidatie van pyriet leidt tot een gelijktijdige verhoging van de Fe^{2+} concentratie en de zuurgraad in de bodem van de schorren en mangroven langs de kust, de oxidatie van hematiet in tropische savanne dalgronden Fe^{2+} verlaagt, maar ook de zuurgraad in het droge seizoen verhoogt. Als een gevolg werd gevonden dat het enkel seizoen irrigatie systeem van Moussodougou significant (p-waarde 0,4%) een aanzienlijk hogere zuurgraad heeft (*pH* 5,7) dan het dubbel seizoen systeem van Tiefora (*pH* 6,4) met ook 750 - 1.800 mg/l hoger ferro-ionen Fe^{2+}.

Het ferro-ijzer gehalte bereikte 3000 mg/l in een aantal lagen in Moussodougou. Dit resultaat is een rechtvaardiging voor het moderniseren van traditionele enkel seizoen vloed irrigatie systemen tot dubbel seizoen geïrrigeerde rijst systemen.

Type ondergrondse drainage hangt af van de verdeling van de klei

Drassige dalgronden in de tropische savanne zijn gebieden waar de rijkste traditionele teelt systemen worden gevonden, maar zij worden ook geconfronteerd met nadelige fysisch-chemische omstandigheden die drastische opbrengst reducties van rijst teweeg kunnen brengen. Ondergrondse drainage wordt in vele gebieden toegepast ter voorkoming van wateroverlast. Deze drainage is echter afhankelijk van de verdeling van het type en de locatie van de klei. Het huidige onderzoek analyseerde deze factoren voor het Tiefora gebied. Hiertoe zijn negen boorgaten, met diepten van 2 tot 6 m, gemaakt. Ongeveer 50 monsters van de bodem zijn op verschillende diepten, op basis van bodem veranderingen in textuur en kleur, genomen. Op deze monsters is korrelgrootte analyse toegepast. Een vergelijkende niet-lineaire regressie is op de verdeling van de klei toegepast. Kwadratische regressie was het meest geschikt. De klei gehalten waren hoog - 20 - 30% in de 2 m bovengrond - in de bovenstroomse en middelste gebieden. Een belangrijker - 30 - 40% - piek werd in het stroomafwaartse gebied op 1 m diepte gevonden, met een veel kleinere dikte (minder dan 50 cm) en hogere doorlatendheid. Deze resultaten suggereren dat de toepassing van mol drainage in de dalgrond, behalve in het benedenstroomse deel waar de klassieke Hooghoudt ondergrondse buizen drainage kan worden aangelegd.

Kosten van ondergrondse drainage kunnen worden beperkt door rekening te houden met de verdeling van de doorlatendheid in de dalgrond

In overstroombare tropische savanne dalgronden resulteren zeer lage infiltratie waarden vaak in zure omstandigheden die gunstig zijn voor hoge concentraties aan metaal ionen die toxisch zijn voor rijst. Bepaling van de infiltratie is van belang voor het ontwerpen van drainage om gedegradeerde bodems te ontginnen. Verschillende studies zijn gericht geweest op het in kaart brengen van de infiltratie waarden. Maar de relatie met de topografie van de vallei is niet verklaard. In dit onderzoek is deze mogelijkheid geanalyseerd door het geval van de dalgrond met geïrrigeerde rijst in Tiefora te bestuderen. Negen boorgaten - 1 tot 5 m diep - zijn van bovenstrooms naar benedenstrooms uitgevoerd. De Lefranc doorlatendheid test voor freatische omstandigheden in dalgronden - te gebruiken wanneer de ondoorlatende laag dicht bij de oppervlakte ligt of afwezig is - is uitgevoerd. Eerst is een vergelijkende regressie toegepast op de gegevens, met inbegrip van alle parameters van de regressie krommen. Bij verschillen in de infiltratie processen was de vergelijking gericht op de uiteindelijke doorlatendheid. Onze resultaten vertonen een toename in de doorlatendheid van bovenstrooms (0,10 ± 0,10 cm/uur) naar benedenstrooms (groter dan 20.0 ± 10.0 cm/uur op sommige plaatsen). Wanneer bij het ontwerp van het ondergrondse drainagesysteem rekening wordt gehouden met een dergelijke toename in de doorlatendheid kan dit leiden tot de toepassing van efficiënter en meer kosten effectieve systemen.

Op gegevens gebaseerd waterbeheer kan helpen om waterverlies te verminderen en fricties tussen boeren ten gevolge van water ongelijkheid op te lossen

Oppervlakte irrigatie vertegenwoordigt meer dan 99% van het geïrrigeerde gebied in West-Afrika en omvat in de regel dalgronden waar geïrrigeerde productie van rijst plaatsvindt, die vaak worden gehekeld als water verspillende systemen. Verrassend genoeg is er geen vervolg noch een analyse van het irrigatiewater gebruik in deze zwaartekracht irrigatie systemen. Een dergelijk onderzoek is uitgevoerd in het 16 ha grote tropische savanne dalgrond systeem met geïrrigeerde rijst in Tiefora. Met de stromingsvergelijking voor de betonnen stuw van het inlaatwerk, werden dagelijks wateraanvoer volumes berekend voor een tijdreeks over een periode meer dan een jaar. De trend analyse van het voortschrijdend gemiddelde laat zien dat zowel tijdens het regen seizoen (1200 mm neerslag) en het droge seizoen (geen regen), de schuif voor het hoofdkanaal bijna nooit gesloten wordt, waardoor een minimale afvoer van 200 m^3/dag voor 4 ha (50 mm/dag versus plaatselijke verdamping van 7 mm/dag) wordt gehandhaafd. Dit benadrukt de noodzaak van een verbeterd waterbeheer. Bovendien toonde de auto correlatie analyse met behulp van het ARIMA model aan dat een irrigatie cyclus die zorgt voor gelijkheid in de verdeling van water onder de percelen 20 dagen is in plaats van vijf. De kennis van dit feit kan potentiële conflicten over gelijkheid onder de boeren verminderen: het gebrek aan water in de 4e dag kan later worden gecompenseerd tijdens de 20-daagse cyclus. Het bleek dat een eenvoudige water meter - geïnstalleerd in het inlaatwerk van het hoofd irrigatie kanaal - een tijdreeks kan opleveren waarbij een autoregressief model voor het voortschrijdend gemiddelde, tegen lage kosten, kan worden toegepast om een gedegen benadering van het waterbeheer in dit oppervlakte irrigatie systeem te realiseren.

Ondergrondse drainage vermindert ijzer toxiciteit op de middenlange en lange termijn

IJzer toxiciteit is een van de belangrijkste beperkingen die de productiviteit van rijst in geïrrigeerde dalgronden in de tropische savanne belemmeren, maar gelukkig kan dat worden verminderd. Een te hoog ferro-ijzer gehalte in de bodem kan de rijstopbrengst teniet doen. Verschillende onderzoeksgebieden - agronomie, bodemkunde via microbiologie - streven ernaar om een oplossing voor dit probleem te bieden. Tot op heden bleef de bijdrage van vloeistofmechanica aan de aanpak van ijzer toxiciteit beperkt. Het huidige onderzoek was gericht op dit aspect door middel van gecontroleerde experimenten met rijst op zeer met ferro-ijzer verontreinigde hematiete grond. In twaalf betonnen microplots en emmers werden twee onafhankelijke experimenten gedurende een periode van 86 dagen uitgevoerd. Drainage en toedienen van kalk waren de twee factoren waarvan de effecten zijn onderzocht. Twee methoden van drainage werden toegepast: 0 mm/dag en 10 mm/dag, het toedienen van kalk vond ook op twee manieren plaats: 0 kg/m² en 1 kg/m² per eenheid toename van de *pH*. Vier verschillende reacties in de bodem zijn gemeten: ferro-ion concentratie Fe^{2+}, *pH*, oxidatie reductie potentieel, en opgeloste zuurstof. Voor de rijst, zijn de toxiciteit scores van het Internationale Rijst Onderzoek Instituut (IRRI) gevolgd. De resultaten wijzen op een toename van Fe^{2+} van 935 mg/l tot meer dan 1106 mg/l (bij een 95% betrouwbaarheid niveau), maar, en dat is interessant, met een significante daling van de zuurgraad van *pH* 5,6 tot 7,3 (95% betrouwbaarheid niveau). Het toedienen van kalk had hetzelfde effect op het verminderen van de zuurgraad. Reductie processen werden niet gehinderd door ondergrondse drainage omdat het oxidatie reductie potentieel daalde van 84,6 tot 9,2 mV, en de opgeloste zuurstof verliep van 1 mg/l tot minder dan 0,1

mg/l. Ondanks de vermindering van de zuurgraad, met zo'n hoog ferro-ijzer gehalte als
1106 mg/l, bereikte de ijzer toxiciteit een score van 7 in de twaalf microplots en de rijst
ging dood. Toch zorgde de verlaging van de zuurgraad van de bodem voor een nieuw
inzicht in het gedrag van de hematiete gronden, tegenovergesteld aan de verzuring door
ondergrondse drainage van pyriet in schorren langs de kust en mangroven. Bovendien
zal het leiden tot minder ferro-ijzer inname door de wortels van de rijst en in een
dergelijk perspectief het verhogen van de rijst opbrengst. Tenslotte, hoewel het
toedienen van kalk hetzelfde resultaat kan bewerkstelligen, heeft ondergrondse drainage
een voordeel wanneer dit mineraal niet beschikbaar of duur is.

Project resultaten voor de boeren in Tiefora

De onderzoeken en hun ondersteunende activiteiten, resulteerden in twee belangrijke
voordelen voor de boeren in Tiefora. Ten eerste, om ijzer toxiciteit te verminderen - die
in dit gebied veel minder ernstig is dan in Moussodougou - en de rijstopbrengst (minder
dan 4 ton/ha) te verbeteren, zou het noodzakelijk zijn om de samengestelde meststof
NPK toe te passen volgens de normen van het Instituut voor Milieu en
Landbouwkundig Onderzoek (IN.ERA). Deze toepassing moet echter samengaan met
het maken van goede grondruggen rond de percelen om de hoeveelheid mest te
beperken en de mineralen meer beschikbaar te laten zijn voor de wortels van de rijst. Dit
zal het gewas stimuleren en versterken en daarmee de weerstand tegen ijzer toxiciteit.
Ten tweede heeft het project aan de vereniging van boeren in Tiefora drie belangrijke
documenten overgedragen: i) een luchtfoto van de omgeving van de vallei van Tiefora,
met inbegrip van het reservoir, het dorp, de wegen en de geïrrigeerde vallei; ii) een
topografische kaart van de vallei, bedoeld om te helpen bij eventuele technische werken
aan het irrigatie systeem, en iii) een gedetailleerde kaart van de ligging van de percelen,
tesamen met een volledige lijst van de boeren en de grootte van hun boerderij, alsmede
de locatie van ijzer houdende percelen voor hun dagelijkse activiteiten.

Project resultaten voor de boeren in Moussodougou

Op basis van de onderzoeksresultaten en vanwege de ernstige ijzer toxiciteit in
Moussodougou, heeft het project een aantal adviezen verstrekt en een aantal belangrijke
documenten aan de boeren overhandigd. De ferro-ijzer concentratie in de bodem van
Moussodougou kan in een aantal percelen 3000 mg/l bereiken met een zuurgraad zo
ernstig als *pH* 4. Omdat werd gevonden dat het voorkomen van ferro-ijzer in de grond
leidde tot intensivering van de activiteit van ijzer-reducerende bacteriën, en gezien het
positieve conserverende effect van organische stof in het verlichten van de
bodemstructuur, heeft het project geadviseerd om het gebruik ervan te reduceren, maar
om het niet volledig te elimineren. Tegelijkertijd zouden de boeren de samengestelde
meststof NPK zoals in Tiefora moeten gebruiken, volgens de normen van IN.ERA,
maar in combinatie met een zorgvuldige opbouw van grond ruggen om het minerale
element meer beschikbaar te maken voor de rijst. Vanwege het feit dat het huidige
enkele irrigatie seizoen gedurende een jaar in Moussodougou een versterkende factor
voor ijzer toxiciteit is, heeft het project ook bij de vereniging van boeren haar lopende
werkzaamheden met betrekking tot beregening uit grondwater tijdens het droge seizoen
geintroduceerd. Tot slot, heeft het project dezelfde set documenten als in Tiefora aan de
vereniging van boeren overgedragen, maar dan met betrekking tot de dalgrond van
Moussodougou.

Andere sociale effecten

In een ultieme poging om de verworven inzichten in het proces om ijzer toxiciteit te verminderen te delen, heeft dit onderzoeksproject een aantal handige video's geproduceerd en geladen in het sociale medium *YouTube*. De 15 geladen en voor iedereen toegankelijk gemaakte video's, behandelen uiteenlopende onderwerpen zoals hydrometrie, microbiologie, geochemie en het assembleren van kleinschalige waterbesparende irrigatie-apparatuur op dorpsniveau (zonder elektriciteit). Veel van deze video's zijn zeer gewaardeerd door het publiek. Zo is bijvoorbeeld, de video betreffende '*Innovatieve irrigatiesystemen in Sub-Sahara Afrika* (in het Frans)' 300 keer per maand bekeken/gedownload. Ook de video '*Hoe neem je een monster van verstoorde bodem of van grond onder water op verschillende diepten* (in het Engels)' is ongeveer 40 keer per maand bekeken/gedownload. Deze twee video's werden geclassificeerd als 'creatief gemeengoed' vanwege hun hoge potentiële kans op overname voor videoproducties door derden. Daarom wordt verwacht dat het project in de komende maanden of jaren een nog groter sociaal effect zal hebben.

H. Sommaire (Langue Française)

La croissance rapide de la population mondiale, contrastée avec la multiplication des obstacles à l'élaboration de nouveaux systèmes d'irrigation, met un défi particulier sur les systèmes de production agricole humains qui doit être relevé. Les populations de nombreux pays en Asie, Moyen-Orient et en Afrique devraient doubler dans les 50 prochaines années. L'expérience de la révolution verte en Asie - au cours de laquelle 70% de l'augmentation de la production alimentaire a été fourni par l'agriculture irriguée - montre qu'il n'y a pas seulement une nécessité de s'efforcer d'augmenter ces systèmes de production agricole, mais aussi d'améliorer l'efficacité de ceux qui existent. En fait, comme un cas beaucoup plus inquiétant, la production de riz irrigué de bas-fond de la zone de Savane Africaine est loin de donner le rendement espéré. L'un des principaux obstacles à la production est la toxicité ferreuse, conséquence de mauvaises conditions de drainage. Selon Africa Rice, au moins 60% des vallées marécageuses en savane tropicale sont touchés par différents degrés de toxicité ferreuse. Le rendement dans de nombreuses zones tombe à zéro, laissant derrière lui des millions d'agriculteurs déçus et pauvres. Par conséquent, il n'est pas surprenant qu'il existe une forte dynamique de recherche - allant de l'agronomie à la microbiologie - qui vise à proposer une solution pour remédier à la toxicité ferreuse du riz. Étant donné que la prévalence de conditions anoxiques dans le sol combinée avec le développement des bactéries réductrices du fer ont été désignées comme un facteur contributif de base à la toxicité ferreuse, cette recherche a choisi d'investiguer les effets du drainage de subsurface sur la résolution de ce défi.

Deux séries complémentaires d'opérations – réparties au sein deux composantes de projet – mettant l'accent sur cinq questions fondamentales étroitement liées aux facteurs contribuant au développement de la toxicité ferreuse ont été menées. En fait, le projet de recherche a été mis en œuvre selon deux composantes: les enquêtes de terrain et les expériences contrôlées. Les enquêtes sur le terrain ont étudié les facteurs déclenchant ou aggravant de la toxicité ferreuse, tels que les proportions d'argile, les ions ferreux Fe^{2+}, la concentration d'oxygène dissous, l'acidité du sol ou de la gestion de l'eau. Tirant profit des connaissances acquises dans ces investigations de terrain et de l'examen de la littérature, deux expériences contrôlées ont été conçues et menées en parallèle. L'une utilisa des microparcelles béton et l'autre des seaux, afin de déterminer statistiquement l'impact des facteurs précédents sur le sol en termes de variation de l'acidité et de la concentration de fer ferreux. Toutes les opérations effectuées dans les deux composantes de ce projet de recherche se sont efforcés de répondre aux cinq questions de recherche suivantes:

1) comment le fer ferreux est-il formé et réparti dans les sols de bas-fond ?

2) comment l'argile est-elle répartie dans le bas-fond?

3) comment la répartition de l'argile dans le bas-fond affecte-t-elle la perméabilité ?

4) comment la gestion de l'eau peut-elle aider à améliorer les conditions de drainage du sol?

5) quel est l'impact de drainage se subsurface sur la toxicité ferreuse ?

Les réponses à ces questions de recherche - déjà publiés ou sous presse - sont exposées ci-dessous, suivies par la contribution de ce projet de recherche dans deux domaines: i) la science et l'ingénierie, et ii) la socio-économie.

L'argile et les ions ferreux peuvent être stratifiés dans les bas-fonds

De fortes concentrations en ion ferreux Fe^{2+}, insérées dans des couches denses en 'argile, constituent une menace importante pour la production de riz dans de nombreux bas-fonds de riz irrigué de la Savane Ouest Africain. Ceci explique pourquoi de nombreuses actions sont actuellement menées pour lutter contre la toxicité ferreuse du riz. Dans cette étude, nous avons investigué la stratification de l'argile et du fer ferreux dans un bas-fond typique de la zone tropicale de savane appelé Tiefora, au Burkina Faso. Prenant en compte les variations de pentes dans le bas-fond, deux séries de prélèvements (pour l'argile et pour le fer) aléatoires de sol ont été effectuées à différentes profondeurs. Les échantillons ont été prélevés à une profondeur de 500 cm, mais surtout à 30 cm, 50 cm et 100 cm. Le pourcentage d'argile a été déterminé par une analyse granulométrique. Les concentrations en fer ferreux ont été obtenues par la méthode réflectométrique. Les stratifications d'argile et d'ions ferreux Fe^{2+} ont été vérifiées à l'aide des tests d'hypothèses statistiques (ANOVA et test t de Welch). Le pourcentage d'argile dans les premiers 100 cm de sol – valant 28,9% – a été trouvé deux fois plus élevé que dans les couches sous-jacentes. En outre, le fer ferreux est principalement situé dans les premiers 30 cm, avec une concentration moyenne de 994 mg / l. Cette concentration en fer ferreux est beaucoup plus élevée que celle observée à une profondeur de 50 et 100 cm en dessous, où la valeur chute à 73 (mg / l), tandis que le pH de l'ensemble des trois couches est pratiquement neutre. Cette stratification frappante suggère plusieurs moyens de réduire la toxicité ferreuse. Parmi ces moyens, nous proposons de maintenir des conditions humides pendant la période de croissance du riz dans les bas-fonds irrigués, en combinaison avec un lessivage par drainage de subsurface dans les périodes de jachère.

Le risque de toxicité ferreuse est plus élevé dans les périmètres irrigués pratiquant une campagne unique

Dans le but de trouver les différences géochimiques et d'aider à construire des stratégies pour lutter contre la toxicité ferreuse, deux bas-fonds de riz irrigué sur des sols riches en hématite ont été étudiés dans la zone tropicale de savane du Burkina Faso. Le premier site était Tiefora, un système moderne 16 ha avec double-campagne de riz irrigué et modérément affecté par la toxicité ferreuse (10% de la surface avec un score de toxicité de 4). Le deuxième site était Moussodougou, un bas-fond traditionnel de 35 ha effectuant seule campagne de riz irrigué, plus sévèrement frappé par la toxicité ferreuse avec plus de 50% des terres affectées (score de toxicité 7). Neuf extraits de sol ont été prélevés à trois profondeurs - 30, 50 et 100 cm - soit 27 à Tiefora et 27 à Moussodogou. Cinq méthodes ont été utilisées pour mesurer les données: i) la concentration en fer ferreux a été déterminée à l'aide d'un réflectomètre, ii) un pH-mètre a donné le pH, iii) les proportions d'argile ont été obtenues par la méthode l'United States Department of Army (USDA) d'analyse granulométrique et de densitométrie, iv) la matière organique a été déterminée par séchage au four et v) la masse volumique apparente à sec a été déterminée en utilisant des échantillons de sol non perturbées. Le test statistique de variance one-way ANOVA et le test t de Welch ont été appliqués aux données pour isoler les similitudes et les différences entre les deux sites. Une analyse géochimique a suivie pour trouver les causes de ces différences. Les résultats ont montré que, à l'opposé de l'oxydation de la pyrite – dans les sols des plaines côtières inondables et les mangroves – qui conduit à une augmentation simultanée des concentrations de Fe^{2+} et de l'acidité, l'oxydation de l'hématite dans les bas-fonds de la zone tropicale de savane diminue Fe^{2+}, mais augmente aussi l'acidité *pendant la saison sèche*. En conséquence, il a été constaté que les sols de Moussodougou, un périmètre à campagne unique de riz, est significativement (p-valeur de 0,4%) plus acide (*pH* 5,7) que les sols du système à double campagne de Tiefora (*pH* 6,4). Cette différence se manifeste également avec le fer ferreux de Moussodougou qui est 750-1800 mg / l supérieur à Tiefora. Le fer ferreux atteint 3000 mg / l dans certaines couches de Moussodougou. Ce résultat procure est une justification pour moderniser un bas-fond traditionnel à campagne unique en système irrigué moderne à double campagne de riz.

Le type de drainage de subsurface repose sur le type de distribution de l'argile

Les sols engorgés de bas-fond de la zone tropicale de savane sont des domaines où les plus riches systèmes de culture traditionnels se trouvent, mais qui sont également souvent confrontés à des conditions physico-chimiques nocives qui peuvent réduire à néant le rendement du riz. Le drainage de subsurface a été utilisé dans de nombreux domaines pour réduire l'engorgement. Cependant, ce drainage dépend de la distribution en profondeur de l'argile, de son type et de son emplacement sur la toposequence. La recherche menée a analysé ces facteurs en utilisant le cas de Tiefora, un bas-fond rizicole irrigué de 16 ha, au Burkina Faso. A cet effet neuf trous de tarière, à des profondeurs de 2 à 6 m, ont été réalisés. Quelques 50 échantillons de sols ont été extraits en fonction des changements de texture et de couleur du sol. Ces échantillons ont subi l'analyse granulométrique. Une régression comparative non-linéaire a été réalisée sur la distribution de l'argile. La régression quadratique fut la plus appropriée. En outre, les proportions d'argile étaient élevées – 20-30% dans les deux premiers mètres de sol – dans les zones en amont et intermédiaire du bas-fond. Un maximum plus important – 30-40% d'argile – a été atteint dans la zone aval à 1 m profondeur, avec une épaisseur beaucoup plus faible (moins de 50 cm) et une plus grande perméabilité. Ces résultats suggèrent l'application du drainage par tunnel dans le bas-fond, à l'exception aval, où le drainage de subsurface classique de Hooghoudt par tubes perforés enterrés peut être mis en œuvre.

Le coût de drainage de subsurface peut être réduit en prenant en compte la distribution de la perméabilité dans le bas-fond

En zone de savane tropicale inondée, les infiltrations très faibles des sols de bas-fonds se traduisent souvent par des conditions acides favorables à de fortes concentrations d'ions métalliques, toxiques pour le riz. La détermination de la perméabilité des sols est importante dans la conception des réseaux de drainage pour récupérer les sols dégradés. Plusieurs études ont porté sur la perméabilité dans les bas-fonds. Cependant, sa relation avec la toposéquence n'a pas été abordée. Cette recherche a investigué cet aspect, en examinant le cas du bas-fond de riz irrigué de Tiefora. Neuf trous de forage – de 1 à 5 m de profondeur - ont été mises en œuvre de l'amont à l'aval. Le test de perméabilité Lefranc sous le niveau de nappe phréatique dans les sols engorgés – utilisé lorsque la couche imperméable est près de la surface du sol ou est absente - a été effectué. Tout d'abord, une régression comparative a été appliquée sur les données, en considérant tous les paramètres des courbes de régression. En cas de différence des processus d'infiltration, la comparaison a porté sur la perméabilité finale. Nos résultats montrent une augmentation de la perméabilité de l'amont (0,10 ± 0,10 cm / h) à l'aval (plus de 20,0 ± 10,0 cm / h dans certains endroits). Tenir compte de cette augmentation de la perméabilité dans la conception des systèmes de drainage de subsurface se traduirait par la mise en place de systèmes plus efficaces et plus rentables.

Une gestion de l'eau basée sur les données collectées peut aider à réduire les pertes d'eau et à résoudre les frictions liées à l'inéquité dans la distribution de l'eau entre les agriculteurs

L'irrigation de surface représente plus de 99% de la superficie irriguée en Afrique de l'Ouest et comprend généralement des bas-fonds dédiés à la production rizicole, qui sont souvent dénoncés comme des systèmes de gaspillage d'eau. Étonnamment, il n'y a ni suivi ni analyse de l'eau d'irrigation utilisée dans ces systèmes d'irrigation par gravité. Un tel travail a été effectué dans le cas du bas-fond rizicole irrigué de 16 ha de Tiefora, en zone de savane tropicale. En utilisant l'équation du déversoir transversal en béton placé à l'entrée du canal primaire, les volumes quotidiens d'utilisation de l'eau ont été calculés sous forme de série chronologique couvrant une période de plus de un an. L'analyse de tendance par la méthode des moyennes mobiles révèle que aussi bien pendant la saison des pluies (1200 mm de précipitations) et la saison sèche (pas de pluie), la vanne du canal principal n'est presque jamais fermée, délivrant un débit minimum de 200 m^3/jour pour 4 ha de terre (soient 50 mm/jour, à comparer avec une évapotranspiration locale de 7 mm / jour). Cela souligne la nécessité d'une gestion de l'eau plus rigoureuse. En outre, l'analyse d'autocorrélation en utilisant le modèle ARIMA a montré que le cycle d'irrigation qui assure l'équité dans la distribution de l'eau entre les parcelles agricoles est de 20 jours au lieu des cinq convenus par les agriculteurs. La connaissance de ce fait peut désamorcer des conflits potentiels au sujet de l'équité entre eux: le manque d'eau dans la journée 4 peut être compensé plus tard au cours du cycle de 20 jours. Par ailleurs, cette étude montre également que, en équipant d'un dispositif simple de mesure de niveau d'eau - installé en début du principal canal d'irrigation – on peut produire une série chronologique à laquelle la méthode des moyennes mobile et l'autocorrélation peuvent s'appliquer, à faible coût, pour faire une évaluation approfondie de la gestion de l'eau dans un système d'irrigation de surface.

Le drainage de subsurface atténue la toxicité ferreuse à court et moyen termes

La toxicité ferreuse est l'une des contraintes les plus importantes qui entravent la productivité rizicole dans les bas-fonds irrigués de la zone de savane tropicale, mais qui peut heureusement être remédiée. Un niveau de fer ferreux trop élevé dans le sol peut annuler le rendement du riz. Plusieurs secteurs de recherche - agronomie, pédologie par la microbiologie - s'efforcent d'apporter une solution à ce problème. Jusqu'à ce jour, la contribution de l'hydraulique à la lutte contre la toxicité ferreuse est restée limitée. La recherche actuelle s'est focalisée sur cet aspect, à travers des expérimentations contrôlées sur sols d'hématite très contaminés en fer ferreux. Douze microparcelles en béton armé et huit seaux ont été utilisés pour mettre en œuvre deux plans d'expériences indépendantes pendant une période de 86 jours. Le drainage de subsurface et le chaulage ont été les deux facteurs dont les effets ont été étudiés. Le drainage de subsurface a été utilisé avec deux conditions de traitement: 0 mm/jour et 10 mm/jour. De même, pour le chaulage deux conditions de traitement ont été mises en œuvre: 0 kg/m² et 1 kg/m² pour avoir une incrémentation du *pH* d'une unité. Quatre variable-réponses différentes dans le sol ont été mesurées les essais: la concentration en ions ferreux Fe^{2+}, le *pH*, le potentiel redox, et l'oxygène dissous. Pour le riz, les scores de toxicité de l'International Rice Research Institute ont été suivis. Les résultats indiquent une augmentation de Fe^{2+} de 935 mg/l à plus de 1 106 mg/l (à 95% de niveau de confiance), mais, ce qui est intéressant, avec une diminution de l'importance de l'acidité du sol de *pH* 5.6 à 7.3 (95% de degré de confiance). Le chaulage a eu le même effet dans la réduction de l'acidité. Les processus de réduction n'ont pas été entravés par le drainage de subsurface puisque le potentiel redox a chuté de 84,6 à 9,2 mV, et que le l'oxygène dissous a décru de 1 mg/l à moins de 0,1 mg/l. En dépit de la réduction de l'acidité, avec un niveau aussi élevé de fer ferreux que 1 106 mg/l, le score IRRI de la toxicité fer atteignit 7 dans les douze microparcelles et tout le riz est mort. Toutefois, la réduction d'acidité du sol apporte une nouvelle compréhension sur le comportement des sols d'*hématite*, qui s'oppose à l'acidification consécutive au drainage de subsurface constatée dans les plaines inondables côtières et les mangroves à dominante *pyrite*. En outre, cette diminution de l'acidité conduira à absorption plus en plus amoindrie de fer ferreux par les racines du riz. Dans cette perspective, le drainage de subsurface permettra d'améliorer le rendement du riz à moyen et long termes. Enfin, si le chaulage peut obtenir le même résultat, le drainage souterrain prend l'avantage quand ce minéral n'est pas disponible ou est cher.

Les résultats du projet pour les agriculteurs Tiefora

A partir des investigations de terrain et les essais en milieu contrôlé, deux avantages majeurs ont été apportés aux agriculteurs de Tiefora. Tout d'abord, afin d'atténuer la toxicité ferreuse - ce qui est beaucoup moins sévère en ce lieu qu'à Moussodougou - et améliorer le rendement du riz (moins de 4 tonnes/ha), il serait indispensable d'appliquer selon les normes de l'Institut de environnement et de la recherche agricole (IN.ERA), l'engrais complexe NPK. Cependant, cette application devrait aller de pair avec la réalisation des diguettes bien construites autour des parcelles agricoles afin de confiner l'engrais et de rendre les minéraux plus disponible pour les racines du riz. Cela fortifierait le riz et renforcerait sa résistance à la toxicité ferreuse. Deuxièmement, le projet a remis à l'association des agriculteurs de Tiefora trois documents clés: i) une photo aérienne de l'environnement de la vallée de Tiefora, y compris le réservoir, le village, les routes et le bas-fond irrigué, ii) une carte topographique du bas-fond, destiné à aider dans les travaux potentiels d'ingénierie sur le système d'irrigation, et iii) une carte détaillée du parcellaire agricole, accompagné de la liste complète des agriculteurs et de leurs tailles d'exploitation, de l'emplacement des parcelles avec indication des zones à toxicité ferreuse pour aider dans les opération quotidiennes.

Les résultats du projet pour les agriculteurs Moussodougou

S'appuyant sur les résultats des investigations de terrain et les expérimentations en milieu contrôlé sur la toxicité ferreuse à Moussodougou, le projet a fourni plusieurs conseils et remis certains documents essentiels pour les agriculteurs. La concentration de fer ferreux dans le sol de Moussodougou peut atteindre 3000 mg/l dans de nombreuses parcelles agricoles avec une acidité aussi grave que *pH* 4. Étant donné qu'on a découvert que l'incorporation de la matière organique dans le sol induisait la croissance des bactéries réductrices du fer, mais à l'opposé étant donné son effet conservatif sur le sol, le projet a avisé les agriculteurs de ne pas complètement abandonner l'utilisation de cette matière organique. Cependant, en parallèle, les agriculteurs devront utiliser l'engrais NPK tout comme à Tiefora, i.c. selon les normes de l'Institut de environnement et de la recherche agricole (IN.ERA), mais aussi en combinant avec une érection de diguettes de parcelles pour confiner l'engrais et le rendre plus disponible pour le riz. En raison du fait que la campagne unique actuelle d'irrigation pendant l'année Moussodougou est un facteur aggravant de la toxicité ferreuse, le projet a également présenté à l'association des agriculteurs son travail en cours qui propose d'installer un équipement d'irrigation par aspersion à partir des eaux souterraines pendant la saison sèche. Enfin, le projet remis à l'association des agriculteurs le même ensemble de documents qu'à Tiefora, mais concernant bien entendu le bas-fond de Moussodougou.

Autres impacts sociaux

Dans un ultime effort pour partager les connaissances acquises sur le processus de réduction de la toxicité ferreuse, ce projet de recherche a produit et mis en ligne plusieurs vidéos utiles sur le réseau social YouTube. Les 15 vidéos placés en ligne et accessibles pour tout le monde touchent des domaines aussi variés que l'hydrométrie, la microbiologie, la géochimie et les équipements économes en eau pour les petits systèmes irrigués que l'on peut facilement assembler au niveau du village (sans électricité). Beaucoup de ces vidéos ont été très appréciés par le public. Par exemple, la vidéo « Systèmes d'irrigation innovants en Afrique sub-saharienne (français) » a été lu et/ou téléchargé plus de 300 fois/ mois à ce jour. De même, la vidéo «Comment prendre un échantillon de sol perturbé ou non dans un sol immergé à des profondeurs différentes (en anglais) », a été consulté et ou téléchargé quelques 40 fois/mois à ce jour. Ces deux vidéos ont été classées «créatif commun» en raison de leur haut potentiel d'appropriation par les productions vidéo tiers. Par conséquent, il est prévu que le projet aura un impact social encore plus élevé dans les mois ou les années à venir.

I.   Sourouma (Bamanan Kan)

Mogow tiayali téliya, ani sonni foro kura labèni geléya, o nana balo sòrò kènè kuraw dililali kè kòròbò bara yé mogow bolo. Adadew bè na kafo fila yé jamana caman na Azi, Araboula ani Farafina san bidourou na ta kònò. Famounia li mi sòrò là « relovution ŋukudjima » wakati là – o mi kònò suman tiaya fan wolonfila tan kònò o bòra sonni forow kònò na là – o ya yira ko ciaman farali sonni forow kan dama tè labòkè, nka foro korow minou bè bara là bi o lou fènè ka kan ka fanga tigui ya. Ka misaliya, malo sènè yòrò minu bè fara kònò law là, o lou ka sòrò ani sènèkè law djigui yé min yé o cè ka dian kòsobè. Malo sòrò ba Savani Tropicali fara kònò o diuku ba do yé négè baga dé yé, o min diu yé foro dougoukolo dji la tèmè géléya yé. Barakè da min toko Africa Rice, ko i na fo i mana fara kònò là kèmè dan Farafina, i ba soro biworo ni nègè baga bè niokon na. Malo bè bè halaki o yòrò caman là, wa o ko kè lén bè sababu yé ka sénékéla millions caman dusukasi ani ka ou fantan ya. O dé y a kè mògò dabali tè ban ka yé ko ŋinili ba bè sén na – ka ta agronomi là ka sé microbiloji ma – walasa ka nègè baga kèlè. Katukuni finè dèsè dugukolo konona la ani nègè dogoya bactériw dé bè na ni nègè baga bana yé malo kan, anw ka ŋinili bara yé i kun sí dugu kolo kono là dji là tèmè ma walissa ka nègè baga kèlè.

J.   About the author

Mister Keïta Amadou was born in Bamako in September 27, 1964. He followed his primary education in Mali at the Lycée Askia Mohamed. He obtained his engineering degree in Civil Engineering/Hydraulic Option at the National School for Engineers (ENI) in Bamako in 1987.

Mister Keïta specialised in Agricultural Engineering at the Ecole Inter-Etats d'Ingénieurs de l'Equipement Rural (EIER, now 2iE) in year 1991. After that, he joined in 1992 the International Water Management Institute (IWMI) in the hydraulic section that implemented a project on irrigation performance assessment and diagnosis in Burkina Faso and Niger.

From 1997 to 2006, Mr. Keïta became the subregional coordinator of the FAO project GCP/RAF/340/JPN, the activities of which were focused on the development and experimentation of sustainable low-cost and water efficient small irrigation systems. The irrigation systems were designed to use simultaneously surface water and shallow groundwater for small scale farming. Three countries were addressed by the activities: Burkina Faso, Mali and Niger. The office of the FAO project was located within the International School for Water and Environmental Engineering (2iE) in Ouagadougou, where Mr. Keïta started giving lectures in irrigation and drainage.

While being coordinator of the FAO project, Mr. Keïta obtained a Bachelor in Sciences Physiques (physics and chemistry) at the University of Ouagadougou in Burkina Faso in 2003.

In year 2008, Mr. Keïta obtained a Master of Science degree at UNESC-IHE Institute for Water Education in Delft, The Netherlands. Afterwards, he returned to 2iE and continued lecturing in irrigation and drainage. He started a PhD research at UNESCO-IHE in 2010.

Mr. Keïta's current research addresses the issue of drainage of crop land for production improvement. Iron toxicity is one of the most important challenges to rice research and production in Africa, and subsurface drainage of waterlogged valley bottom irrigated rice systems is viewed as a viable solution to alleviate iron toxicity in Tropical Savannah valley bottom soils.

Netherlands Research School for the
Socio-Economic and Natural Sciences of the Environment

D I P L O M A

For specialised PhD training

The Netherlands Research School for the
Socio-Economic and Natural Sciences of the Environment
(SENSE) declares that

Amadou Keïta

born on 27 September 1964 in Bamako, Mali

has successfully fulfilled all requirements of the
Educational Programme of SENSE.

Delft, 26 March 2015

the Chairman of the SENSE board

Prof. dr. Huub Rijnaarts

the SENSE Director of Education

Dr. Ad van Dommelen

The SENSE Research School has been accredited by the Royal Netherlands Academy of Arts and Sciences (KNAW)

K O N I N K L I J K E   N E D E R L A N D S E
A K A D E M I E   V A N   W E T E N S C H A P P E N

The SENSE Research School declares that Mr Amadou Keïta has successfully fulfilled all requirements of the Educational PhD Programme of SENSE with a work load of 39 EC, including the following activities:

SENSE PhD Courses

○ Environmental Research in Context (2010)
○ Research in Context Activity: 'Co-organising the 7th Edition of 2iE Scientific Days Workshop', Ouagadougou, Burkina Faso (2013)

Other PhD and Advanced MSc Courses

○ Summer School 'World History of Water Management', UNESCO-IHE, Delft (2010)

Management and Didactic Skills Training

○ Lecturing in MSc courses 'Localised irrigation' (2008-2014), 'Drainage of agricultural lands (in French)' (2009-2012), and 'Broad area irrigation (pivot systems)' (2009-2014), International Institute for Water and Environmental Engineering (2iE), Ouagadougou, Burkina Faso
○ Lecturing in BSc course 'Sprinkler irrigation (in French), International Institute for Water and Environmental Engineering (2iE), Ouagadougou, Burkina Faso (2009-2013)
○ Co-organising the Africa Water Forum 2014, Ouagadougou, Burkina Faso (2014)

Oral Presentations

○ *Characterisation of the profile and Water circulation in valley bottom rice irrigated soils: case of Tiefora in Burkina Faso*. 11th ICID International Drainage Workshop on Agricultural Drainage Needs and Future Priorities, 23-27 September 2012, Cairo, Egypt
○ *Rice yield improving factors in tropical savannah of Burkina Faso*. UNESCO-IHE - PhD Week, 1-5 October 2012, Delft, The Netherlands
○ *Significant gap in ferrous iron concentration and pH in two rice irrigated valley bottoms soils affected by iron toxicity*. UNESCO-IHE - PhD Symposium, 23-24 September 2013, Delft The Netherlands
○ *Assessing Irrigation Water Management Using Trend Analysis and Autocorrelation*. The 18th World Congress of CIGR, 17-19 September 2014, Beijing, China

SENSE Coordinator PhD Education

Dr. ing. Monique Gulickx

T - #0391 - 101024 - C48 - 244/170/15 - PB - 9781138028166 - Gloss Lamination